ISBN 978-0-266-89985-3
PIBN 10906728

This book is a reproduction of an important historical work. Forgotten Books uses
state-of-the-art technology to digitally reconstruct the work, preserving the original format
whilst repairing imperfections present in the aged copy. In rare cases, an imperfection in
the original, such as a blemish or missing page, may be replicated in our edition. We do,
however, repair the vast majority of imperfections successfully; any imperfections that
remain are intentionally left to preserve the state of such historical works.

TABLE OF SPECTRA ACCORDING TO KIRCHOFF & BUNSEN.

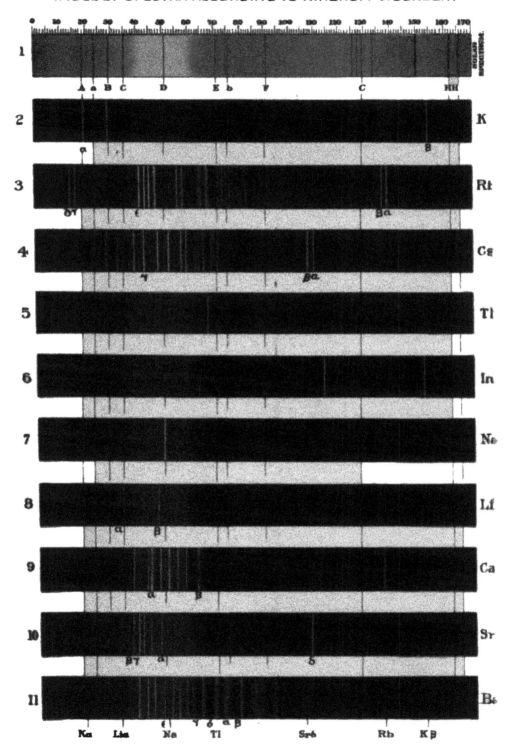

ELDERHORST'S MANUAL

REWRITTEN AND REVISED.

MANUAL

OF

QUALITATIVE

BLOWPIPE ANALYSIS,

AND

DETERMINATIVE MINERALOGY.

BY

HENRY B. NASON,

PROFESSOR OF CHEMISTRY AND MINERALOGY,
RENSSELAER POLYTECHNIC INSTITUTE,
TROY, N. Y.

PHILADELPHIA:
PORTER & COATES.

Copyright,
PORTER & COATES.
1881.

PREFACE.

A MANUAL of blowpipe analysis was prepared by Prof. Elderhorst nearly twenty-five years ago for use in the Rensselaer Polytechnic Institute, there being at that time no text-book which seemed well adapted to the course of instruction as it was here given. To the second edition a translation of a portion of von Kobell's *Tables* (fifth edition) was added.

A complete revision, with many alterations and additions, was afterward made by the editor of this volume, with the assistance of Prof. C. F. Chandler, which has continued in use to date, having passed through several editions. As the demand for the book has continued, it has been thought advisable to bring it up to the present time, which has involved a total change and nearly an entire rewriting.

A new arrangement of subjects, a change of nomenclature and chemical formulas, have been made, and many new methods of determination of compounds and minerals have been introduced. Still, the present work may be considered as based upon, or having grown out of, the later editions of Elderhorst's *Manual.*

In the present compilation the following works have been consulted :

The editions of Elderhorst's *Manual* revised by Nason and Chandler; the German translation of the same by Landauer; von Kobell's *Tafeln zur Bestimmung der Mineralien* (11th edition); ˙ Rammelsberg's *Mineral-Chemie;* Plattner's *Blowpipe Analysis*, translated by Cornwall; Dana's (J. D.) *Descriptive Mineralogy;* Dana's (E. S.) *Text-Book of Mineralogy;* Ross's *Pyrology;* and various articles in scientific journals.

My thanks are due for many valuable suggestions to Dr. C. F. Chandler of Columbia College, Dr. H. C. Bolton of Trinity College, Prof. A. R. Leeds of the Stevens Institute of Technology, and especially to Dr. J. Landauer of Braunschweig, whose *Tables* have been introduced in this work.

My thanks are also due to Mr. W. P. Mason, C. E., my assistant, for much valuable aid in the preparation of this volume.

WINSLOW LABORATORY,
RENSSELAER POLYTECHNIC INSTITUTE,
Troy, N. Y., Dec. 15, 1880.

INTRODUCTION.

AT present a knowledge of blowpipe operations is considered not only convenient, but quite indispensable, to the chemist, mineralogist, geologist, and mining engineer. The small amount of apparatus and few reagents necessary, the ease, quickness, and certainty with which the results are obtained, often render blowpipe methods preferable to all others.

A list of the more important apparatus and reagents is given in Chap. I.

The structure of the flame, and the manner of producing the oxidizing and reducing flames, are explained at the beginning of Chap. II.

In accordance with the instructions there given, with a little practice these flames may be readily produced, and also a steady and continuous blast.

A series of experiments may then be made with suitable substances "In the closed tube," "In the open tube," etc.

The following substances will be found well adapted for exhibiting the effects which may be produced:

1. For examination in the glass tube closed at one end: lime, ammonium nitrate, zinc oxide, mercury oxide, copper sulphate, lead carbonate, manganese dioxide, stibnite, pyrite, cinnabar, siderite, fluorite (see pars. 8–12).

2. In the open tube: bismuth, selenium, arsenopyrite, pyrite, stibnite, cinnabar, galenite, sphalerite, molybdenite (see pars. 13–20).

3. On charcoal or aluminium foil: arsenopyrite, selenium, pyrite, antimony, bismuth, zinc, lead, potassium chlorate, sodium carbonate (see pars. 21–38).

4. In the borax or salt of phosphorus bead: cobalt, nitrate, iron oxide, manganese dioxide, chromium oxide, copper oxide, silica (see pars. 39–41).

1•

5. With sodium carbonate : metallic oxides, silica, sulphur, barium, and strontium (see pars. 42–46).

6. With sodium thiosulphate : metallic oxides (see par. 47).

7. With acid potassium sulphate : nitrates, chlorates, iodides, bromides, acetates, etc. (see pars. 48–50).

8. With zinc and hydrochloric acid after previous decomposition : molybdenum oxide, etc. (see par. 51).

9. With cobalt solution : alumina, magnesia, zinc oxide, etc. (see pars. 52–55).

10. For flame coloration : sodium, potassium, barium, strontium, etc. (see pars. 56–61).

The following substances may then be examined, with and without fluxes, as indicated in the tables on pages 189–215 :

1. Iron sesquioxide, all the reactions given in Table II. 13.
2. Manganese dioxide, Table II. 16.
3. Chromium sesquioxide, Table II. 6.
4. Cobalt and nickel oxides, Table II. 7, 19.
5. Copper oxide, Table II. 8.
6. Zinc oxide, Table II. 35, and metallic zinc (pars. 34–55).
7. Tin oxide, Table II. 30, and metallic tin (par. 28).
8. Lead oxide, Table II. 15, and metallic lead (par. 27).
9. Bismuth oxide, Table II. 3, and metallic bismuth (pars. 17–26).
10. Antimony trioxide, Table II. 1, and metallic antimony (pars. 16–25).
11. Arsenic trioxide, Table II. 2 (pars. 15–33).
12. Mercury oxide, Table II. 17.
13. Alumina, Table I. 5, and par. 55.
14. Magnesia, Table I. 4, and par. 55.
15. Silica (par. 55).
16. A sulphide (pars. 10, 14, 121).
17. A borate (par. 75).
18. A chloride (pars. 82, 83).

Having made these examinations carefully, and having performed all the operations indicated, the analyses of simple substances may be undertaken, and afterward of the more complex and difficult.

Methods of experimenting may be learned from the examples given, or, if considered preferable, the analyses may be made by use of the systematic tables given on pages 169–

188. Such tables may be of some assistance to the beginner, but after little experience it will be better to lay them aside and depend upon the judgment in regards to the tests to be made and the inferences to be drawn therefrom. This method may be illustrated and easily understood from the following examples given by Elderhorst:

1. The substance under examination is antimony sulphide.

Examination in a matrass: At a very high temperature a black sublimate is obtained, becoming reddish-brown when cold. In reading over the list (in par. 10) we find this character belonging to antimony sulphide.

Examination in an open glass tube: Gives sulphur dioxide, detected by the odor and action on blue litmus-paper, and white fumes, which partly condense in the tube. On examining the sublimate with a magnifying-glass, it is found to be amorphous, hence must be antimony trioxide (par. 16).

Examination on charcoal alone: Is completely volatilized, with emission of sulphur dioxide, and deposits a white volatile coating, possessing the properties of the coating of antimony (par. 25).

These few operations are quite sufficient to establish the nature of the substance under trial, since the absence of the more fixed metals proved by the volatility of the substance on charcoal and in the open tube, and the absence of metals giving coatings by the purity of the antimony coating. The presence of arsenic would have been betrayed by an alliaceous odor when heated on charcoal. The only substance which would have escaped detection by these operations is mercury sulphide. In order to ascertain its presence or absence, we perform the operation given under "*Mercury*" in Chap. III.

The result giving an answer in the negative, the body was "antimony sulphide."

2. The substance under examination is lead chromate.

Examination in a matrass: } Fuses and changes color, but
Examination in an open tube: }   gives nothing volatile.

Examination on charcoal alone: Fuses, gives small metallic globules, and deposits a coating which is lemon-yellow while hot, and sulphur-yellow when cold, indicative of lead (par.

8

9

10

11   Lia   Na   Tl   Srb   Rb

indicative of
ins.

the globule
the lead is
e; the borax
of the char-
be well ob-
rom the me-
a wire, and
e; the bead
ulting Table
r oxide, and
e; to decide
action of the
proving the

d to suspect
is beyond a
nd a method
ascertained
ving to deal
with vitrified
fluous. We
he oxidation
motion, the
silver, some
nove the last
obule exhib-

copper, and

nickel, con-

limate, con-
ce of arsenic

ds: Gives a
and a faint
e of sulphur
bere we find

23). It is always desirable to collect the metal to a large globule, and to study its physical properties. This end is best attained by mixing the substance with sodium carbonate and a little borax, and exposing the mixture to the reduction flame on charcoal. In this particular case a metallic button is obtained which is soft, may be flattened by the hammer and cut by the knife—properties belonging to metallic lead.

Examination with borax and salt of phosphorus: Before proceeding with this examination it is necessary to test the substance for the presence of sulphur after the method given in par. 121 (unless the presence of this element was detected by the examination in the open glass tube or on charcoal alone); no sulphur being present, borax and salt of phosphorus beads are made on charcoal, and small portions of the substance added. With both fluxes nearly the same reactions are obtained; in oxidation flame dark-red while hot, and fine yellowish-green when cold; in reduction flame green, hot and cold. In order to find out what body produces such reactions, we use Table II., which leads us to chromium sesquioxide. To corroborate the result, the substance may be fused with sodium carbonate and nitre, as described (par. 85).

The physical properties of the body under trial lead to the final conclusion that it must be lead chromate.

3. The substance is an alloy of silver, copper, and lead.

Examination in a closed tube : } no change.
Examination in an open tube : }

Examination on charcoal alone: Fuses and deposits a copious coating, which is lemon-yellow while hot and sulphur-yellow when cold, indicative of lead (par. 27); the coating cannot contain any bismuth oxide, because the color would be darker in this case, but might contain zinc oxide or antimony oxide. The test is for the presence of the former, the coating is played upon with the oxidation flame: it is completely volatile, hence no zinc present (might also be tested with cobalt solution, par. 54); to test the coating for the presence of antimony oxide, it is scraped off from the charcoal and dissolved in a bead of salt of phosphorus (v. page 56), or the alloy is treated with boric acid, as described under the head of "*Antimony*" in Chap. III. If the blast be continued for a long time, a faint dark-red coating is formed near the assay-piece, indicative of silver (par. 29), and a dark metallic globule remains.

Examination with borax and salt of phosphorus : the globule remaining on the charcoal after volatilization of the lead is treated with borax on charcoal in oxidation flame; the borax becomes colored. Owing to the reducing effect of the charcoal, the influence of the oxidation flame cannot be well observed on charcoal; hence the borax is removed from the metallic globule, fastened into the hook of a platina wire, and here exposed to the action of the oxidation flame; the bead is green while hot, and blue when cold. On consulting Table II. we find that this reaction is produced by copper oxide, and by a mixture of cobalt oxide and iron sesquioxide; to decide between the two, we now expose the bead to the action of the reduction flame; it becomes red and opaque, thus proving the presence of oxide of copper.

By examination on charcoal alone, we were led to suspect the presence of silver; in order to establish this beyond a doubt, we refer to Chap. III., "Silver ;" here we find a method (par. 119) by which the presence of silver may be ascertained in compounds of all descriptions. In our case, having to deal only with lead, copper, and silver, the treatment with vitrified boric acid and metallic lead is, of course, superfluous. We place our alloy at once on the cupel, and direct the oxidation flame upon it; if, after cessation of the rotatory motion, the globule should not possess the bright lustre of silver, some pure metallic lead has to be added in order to remove the last traces of copper. We finally obtained a bright globule exhibiting all the characteristic properties of silver.

Thus we have established the presence of lead, copper, and silver.

4. The substance under examination is copper nickel, containing arsenic, sulphur, nickel, cobalt, and iron.

Examination in a matrass : Gives a slight sublimate, consisting of octahedral crystals, pointing to the presence of arsenic (par. 11).

Examination in a glass tube open at both ends : Gives a copious crystalline sublimate of arsenic trioxide, and a faint odor of sulphur dioxide ; to establish the presence of sulphur beyond doubt, we refer to Chap. III., "Sulphur," where we find

the method (par. 121) for discovering sulphur when in cor
bination with other substances. In performing the test the
described we obtain the sulphur reaction.

Examination on charcoal alone: Gives abundant arsen
fumes, leaving a metallic globule which, even with continue
blowing, does not give rise to the formation of a coating c
the charcoal (absence of volatile metals).

Having removed all volatile substances, we now proceed
examine the remaining globule. On applying a magnet, v
find it powerfully attracted, showing the presence of eith
iron, nickel, or cobalt, perhaps all of them, either alone
combined with other non-volatile metals. We add some bora
to the globule and expose it to the action of the oxidation flam
then remove the borax from the globule, fasten it into the hoc
of a platinum wire, and here observe the color: green whi
hot, blue when cold as in the preceding case (example 3), b
on exposing the bead to the action of the reduction flan
(which is best done by placing it on charcoal and touching
with tin), it does not become brown and opaque, showing ther
fore the presence of a small quantity of iron with cobalt. W
now add a fresh portion of borax to the metallic globule,
order to see whether it consists entirely of cobalt (that it ca
not contain any considerable amount of iron is proved by tl
appearance of the cobalt reaction in the first trial, iron beir
much more readily dissolved by borax than cobalt): the bea
is violet while hot, and assumes a brownish color on cooling
by referring to Table II. we see that this effect is produce
by nickel containing cobalt. Referring to Chap. III., "Nickel
we find the method to detect the presence of this metal whe
in combination with iron and cobalt, and also the presence c
copper, if the assay should contain a small quantity of it.

By the above examples the use of the methods given in tl
third chapter will be sufficiently illustrated. If the substan
under examination be of a simple composition, its nature
readily ascertained by following the general method laid dov
in the second chapter; but if the reactions obtained clear
point to the complex nature of the body, we refer to the r
spective sections of Chap. III.; if, for example, we suspect tl
presence of cobalt in a mineral consisting of arsenides, v
test the substance according to par. 86; if a small quantity c

copper is to be discovered in a mineral, we proceed as directed in par. 88, etc.

In order to obtain characteristic reactions and satisfactory results it is important to experiment upon a suitable substance. A list of such substances which are sufficient to illustrate all the important reactions may be found on page 16. In Chapters II. and III., when a reaction is mentioned or a process described, a number is often added which refers to the substance in the list well adapted for the experiment, p. 12.

In Part II., Chap. VI., all necessary instructions for the determination of mineral species are given. The names of the most important minerals, or of most frequent occurrence, are printed in large type; of the less common, in somewhat smaller type; and the comparatively rare or wholly foreign, in *italics*.

A set of minerals, in small fragments, placed in corked glass tubes, numbered and arranged in a small box provided with partitioned trays, will be found very convenient for study and for reference. The box and trays may be made of heavy pasteboard, and the following dimensions will answer well the purpose: seventeen centimeters in length, thirteen centimeters in breadth, and four centimeters in depth. The glass tubes are five and a half centimeters in length and one centimeter in diameter. The trays, three in number, will hold seventy-two specimen tubes, in which the minerals of the list on the following page may be placed. These having been carefully studied, but little difficulty will be found in determining any of the minerals described in the tables in Chapter VI.

MINERALS FOR BLOWPIPE ANALYSIS.

Potassium.
Orthoclase.
Apophyllite.

Sodium.
Albite.
Natrolite.
Cryolite.

Lithium.
Lepidolite.
Spodumene.

Barium.
Barite.
Witherite.

Strontium.
Celestite.
Strontianite.

Calcium.
Fluorite.
Gypsum.
Apatite.
Calcite.
Wollastonite.
Prehnite.

Magnesium.
Magnesite.
Talc.

Dolomite.
Serpentine.

Aluminium.
Corundum.
Cyanite.
Topaz.

Glucinum.
Beryl.

Manganese.
Pyrolusite.
Franklinite.
Rhodonite.

Iron.
Hematite.
Limonite.
Magnetite.
Siderite.
Pyrite.

Cobalt.
Cobaltite.

Nickel.
Niccolite.

Zinc.
Zincite.
Sphalerite.

Calamine.
Willemite.

Lead.
Galenite.
Cerussite.
Pyromorphite.
Bournonite.

Tin.
Cassiterite.
Stannite.

Bismuth.
Native bismuth.

Copper.
Cuprite.
Chalcocite.
Bornite.
Chalcopyrite.
Tetrahedrite.
Malachite.
Azurite.

Mercury.
Cinnabar.

Silver.
Argentite.
Cerargyrite.
Pyrargyrite.

Titanium.
Menaccanite.
Rutile.

Antimony.
Stibnite.

Tungsten.
Wolframite.

Molybdenum.
Molybdenite.

Chromium.
Chromite.

Arsenic.
Realgar.
Orpiment.
Arsenopyrite.

Boron.
Sassolite.
Tourmaline.

Silicon.
Quartz.

Carbon.
Succinite.
Graphite.

12

CONTENTS.

PART I.

BLOWPIPE ANALYSIS.

CHAPTER I.

CHAPTER II.

CHAPTER III.

CHAPTER IV.

CHAPTER V.

PART II.

DETERMINATIVE MINERALOGY.

CHAPTER VI.

CHAPTER VII.

ABBREVIATIONS.—R. F. Reducing flame.—O. F. Oxidizing flame.—G. Specific gravity.—H. Hardness.—Aq. Water.—A barred letter signifies two of the element.

LIST OF SUBSTANCES

Well Adapted for Showing the most Important Blowpipe Reactions.

Metals.
1. Antimony.
2. Arsenic.
3. Lead.
4. Bismuth.
5. Cadmium.
6. Zinc.
7. Tin.
8. Silver.

Alloys.
9. Mercury and tin.
10. Lead and antimony.
11. Lead and bismuth.
12. Lead and zinc.
13. Lead, copper, and silver.
14. Copper and zinc.
15. Copper and tin.
16. Zinc and cadmium.

Sulphides.
17. Arsenic and antimony (artificial).
18. Arsenic, antimony, lead, and copper (artificial).

Oxides.
19. Antimony oxide.
20. Bismuth oxide.
21. Cadmium oxide.
22. Zinc oxide.
23. Tin oxide.

24. Iron oxide.
25. Chromium oxide.
26. Copper oxide.
27. Cobalt monoxide.
28. Uranium oxide.
29. Tungsten trioxide.
30. Molybdenum trioxide.
31. Arsenic trioxide.
32. Alumina.

Salts.
33. Borax.
34. Salt of phosphorus.
35. Sodium carbonate.
36. Acid potassium sulphate.
37. Ammonium chloride.
38. Potassium chloride.
39. Potassium bromide.
40. Potassium iodide.
41. Sodium chloride.
42. Potassium chlorate.
43. Lead nitrate.
44. Cobalt nitrate.
45. Nickel oxalate.
46. Copper sulphate.
47. Copper chloride.
48. Copper arsenate.
49. Mercurous chloride.
50. Mercuric chloride.

Minerals.
51. Quartz.
52. Gypsum.

53. Calcite.
54. Strontianite.
55. Witherite.
56. Magnesite.
57. Muscovite.
58. Orthoclase.
59. Albite.
60. Petalite.
61. Hematite.
62. Rutile.
63. Pyrolusite.
64. Lepidolite.
65. Apatite.
66. Franklinite.
67. Uraninite.
68. Chromite.
69. Cerussite.
70. Malachite.
71. Stibnite.
72. Pyrite.
73. Chalcopyrite.
74. Arsenopyrite.
75. Smaltite.
76. Cobaltite.
77. Realgar.
78. Cinnabar.
79. Niccolite.
80. Molybdenite.
81. Berthierite.
82. Bournonite.
83. Tetrahedrite.
84. Tiemannite.
85. Sylvanite.
86. Descloizite.

16

PART I.
BLOWPIPE ANALYSIS.

CHAPTER I.

APPARATUS AND REAGENTS.

1. THE blowpipe in general use at the present time is shown in Fig. 1. It consists of a conical tube, *a b*, provided with a mouth-piece, *a*; a cylindrical chamber, *c d*, to retain the condensed moisture of the breath; and a short tube, *f g*, inserted at right angles to this chamber. This latter tube is terminated with a platinum jet, *h*, which may be adjusted to hold by friction or may be soldered to the tube. The terminal opening of the jet should be 0.4 mm. in diameter. When a stronger flame is required, a diameter of 0.5 mm. will be found useful. Should the opening become obstructed, it may be cleared by removing the jet from the tube and heating to moderate redness. The trumpet-shaped mouth-piece is generally preferred, although other forms may be used.

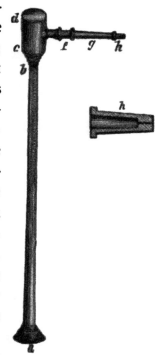

Fig. 1.

The usual length of the blowpipe, without the mouth-piece, is 200 mm., but this length may be varied to suit the visual distance of the operator.

Other forms of blowpipe for special work are frequently employed; as, for instance, the stand blowpipe, which permits both hands being at liberty. This latter, in some cases, has a mechanical blowing-attachment operated by the hand or foot. In most laboratories hydraulic pressure may be conveniently used for producing a steady and long-continued blast. The gas blowpipe, Figs. 2 and 3, supplying the fuel and blast through the

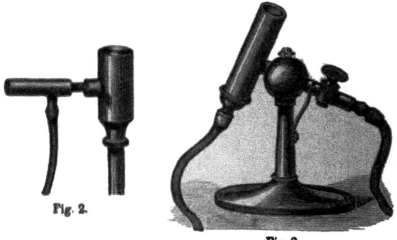

Fig. 2.

Fig. 3.

name jet, and used with or without a mechanical blower, is particularly worthy of notice.

In Fletcher's blowpipe the end of the tube is bent round several times, and this becomes heated by the flame when in use. A hot-air blast is thus produced, and the temperature of the blowpipe flame is considerably increased.

2. Any kind of a flame may be used for the blowpipe, provided it be not too small. The flame of illuminating gas, however, is most convenient for blowpipe experiments, except when testing for sulphur, and the Bunsen burner best suited for the purpose. Fig. 4. The burner rests on a foot, *a b*, into which a block, *c d*, is screwed. To this is attached the tube, *e f*, in which the gas, coming through *k*, mixes with the air drawn through the

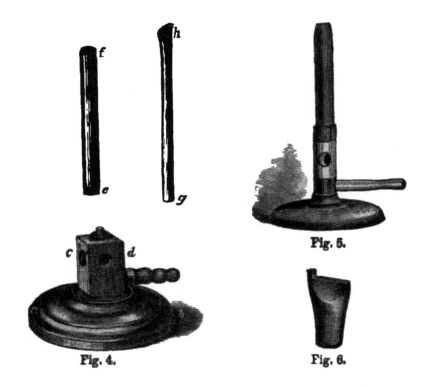

Fig. 4.

Fig. 5.

Fig. 6.

holes in *c d*, and burns at *f* with an almost non-luminous flame.

For blowpipe purposes, the tube *g h*, flattened at the top and slanted, is introduced into the tube *e f*, through which the gas passes without being mixed with the air, and burns with its usual luminous flame. The heating of substances in glass tubes and matrasses is best performed over the non-luminous flame, as it deposits no soot, or over a common alcohol lamp.

An improved form of burner by Morton is shown in Fig. 5, which does not allow the flame to retreat into the tube, but is extinguished under conditions which would cause retreat of the flame in other burners.   A cap for the above burner by Leeds is shown in Fig. 5, which converts the flame into one fit for use with the blowpipe, and furnishes support to steady and direct the blowpipe jet.

The Berzelius blowpipe lamp, Fig. 7, consists of a receptacle for oil, provided with a flat wick, and supported upon a stand. Refined rape-seed or olive oil should be used in this lamp, being cleaner and having less odor than sperm oil.

Fig. 7.

A very convenient form of lamp is described by Fos-

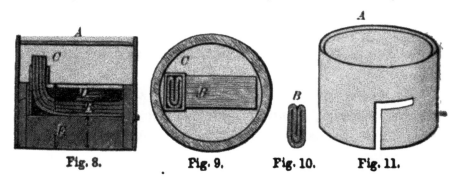

Fig. 8.   Fig. 9.   Fig. 10.   Fig. 11.

ter.   It may be made of tin-plate, brass, or other metal. A vertical section of the whole lamp is shown in Fig. 8.

A is the cover, also shown in Fig. 11, which covers the whole lamp, and is kept in place by a knob and slot. It may also serve as a stand for the lamp when in use. E is the body of the lamp; B the flat-folded wick, shown also in Fig. 10; C the wick-holder, shown also in Fig. 9, a view of the lamp, without cover, from above; D the solid fuel, which may be tallow, stearin, paraffin, or wax, etc.

In using the lamp, the wick is lighted, and the flame then directed upon the fuel so as to melt it.

After use, and before the fuel solidifies, it is well to draw the wick up a little, that it may be ready for lighting at another time.

An alcohol lamp with a flat wick, or an ordinary low "fluid" lamp with a small cylinder, which slides up and down on the tube holding the wick, by which the flame may be adjusted, is in some respects the best substitute for the Bunsen burner and illuminating gas. In this latter form of lamp a mixture of one part of turpentine, or three parts of benzol, to twelve parts of alcohol, should be used.

The heat produced by the flame of a good stearic acid candle is quite sufficient for most blowpipe purposes. To prevent the running of the melted material, Casamajor recommends wrapping the candle closely with rather thick tin-foil. The edges should be folded together several times and then pressed against the candle, so as to form a close joint.

3. **Supports.** Charcoal, platinum, and glass are principally used as supports. Wood charcoal is in most cases the best, on account of its infusibility, non-conductivity, and reducing power. It must be well burnt, and

not scintillate or smoke; it must leave but little ash; charcoal of light-wood, as alder and pine, has been found the best.   It should be sawn into blocks about 10 cm. long and 3 cm. in breadth and thickness. Only those sides which show the rings of growth should be used.

An excellent substitute for wood charcoal is prepared by mixing charcoal dust with starch paste, moulding into the desired form, drying, and heating to dull redness in a closed vessel or until combustible gases cease to be given off.

Aluminium foil has been highly recommended by Ross, in his treatise on "Pyrology," as a substitute for charcoal, and may be often used to advantage, especially for sublimates or coatings.   On to a piece of heavy foil, about 120 mm. long, 35 mm. broad, and one end bent up so as to form a rim 20 mm. deep at a slightly acute angle to the body of the foil, a small piece of charcoal, about 12 mm. square and about the thickness of a penny, is laid, upon which the substance under examination is placed.   The foil may be cleaned by rubbing with a little bone-ash and water, and is quite durable.

Platinum is used whenever the reducing action of the charcoal acts injuriously.   It is advantageously employed on all occasions where no reduction to the metallic state takes place, since the color of the flux is much better seen on the platinum than on charcoal.   It is mostly used in the shape of wire, the end of which is bent so as to form a hook, which serves as a support to the flux.   Fig. 12.   For convenience it is sealed into a drawn-out glass tube, which serves as a handle.   The U-shaped hook forms a spherical bead, and is generally used, whilst the oval-shaped hook forms a flatter bead, and is preferable when the color of the bead is very deep.   A small platinum spoon or capsule, of from about 12 to 15 mm. diameter, is very convenient for fusing substances with fluxes, as nitre, acid potassium sulphate, etc. Figs. 13 and 14. A rectangular piece of **platinum foil**, about 50 mm. long and 15 mm. wide, bent up at the sides and end, may be used in place of the spoon.

Impurities are dissolved and removed from the platinum by placing in dilute sulphuric acid and rinsing with water.

Glass tubes, open at both ends, are used for calcination, and for testing the presence of substances which are volatile at a high temperature. The tubes should be about 6 mm. in diameter and 80 mm. in length. Of glass tubes, sealed at one end, or small **matrasses,** an assortment

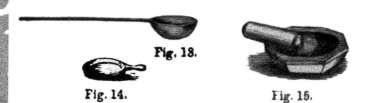

Fig. 12.   Fig. 14.   Fig. 18.   Fig. 15.

should always be kept on hand, since they are of very frequent use.   Figs. 16, 17, 18, 19.

4. Of other apparatus, the most necessary are :

An **agate mortar** 50 to 60 mm. diameter, with pestle of the same material.   Fig. 15.

Forceps with **platinum points.**

Forceps of steel.

A pair of **cutting pliers** for taking small pieces from a mineral specimen.

A small **hammer** and **anvil**, both of steel and well polished.

A three-cornered **file** for cutting glass tubes, trying the hardness of minerals, etc.

A small **magnet**.

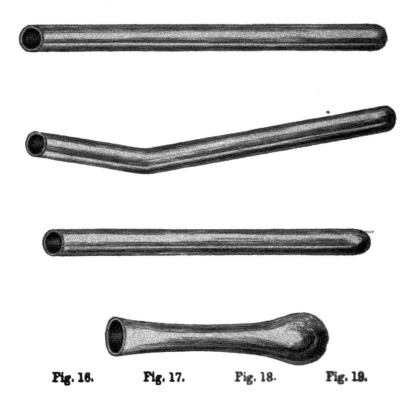

Fig. 16.   Fig. 17.   Fig. 18.   Fig. 19.

A **magnifying-glass** or lens.

Coal-borers. Figs. 20, 21, 22.

A set of **watch-glasses**, which are very convenient for the reception of the assay-piece, the metallic globules, etc.

Pieces of **colored glass**, about 12 cm. long and 5 cm. wide; a blue one, colored with cobalt; a violet, colored with manganese; a red, colored with copper; and a green, with iron and copper.

A hollow **glass prism**, filled with indigo solution. Fig. 23.

This prism is made of plate glass, and filled with a solution prepared by dissolving one part of indigo in eight parts of fuming sulphuric acid, adding fifteen hundred to two thousand parts of water, and filtering. In practice the prism is held close to the eye and moved horizontally, so as to allow the light of the colored flame to reach the eye through successively thicker portions of the absorbing medium.

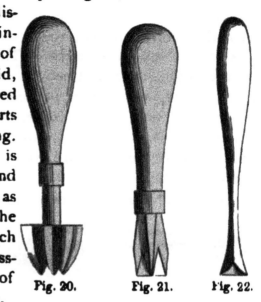

Fig. 20. Fig. 21. Fig. 22.

Another form of prism, somewhat cheaper and more convenient, is shown in Fig. 24.

5. **Reagents.** **Sodium carbonate, borax, and salt of**

Fig. 23.

phosphorus are the most important, but there are others which, though not so extensively used, still are indispensable for the detection of certain substances ; others, the use of which is very limited, are omitted in this list. All should be as pure as possible.

3

Sodium carbonate, Na,CO,. The disodium carbonate or the monosodium, NaHCO$_3$, may be used. It must be perfectly free from sulphuric acid, for the presence of which it may be tested as shown par. 121. It is used as a dissolving, decomposing, and reducing agent.

Borax, Na,B,O, + 10H,O. The commercial article is purified by recrystallization, the crystals washed with distilled water, dried, and reduced to a coarse powder. Borax fuses with intumescence, forming a vitreous mass, which dissolves metallic oxides, showing characteristic colors. **Sodium thiosulphate.**

Salt of phosphorus [microcosmic salt], NH$_4$NaHPO$_4$ + 4H$_2$O. It is used for the same purposes as borax, but

Fig. 24.

sometimes gives more intense or different colors with the metallic oxides. When pure it gives a glass which, on cooling, remains transparent; if this is not the case, it must be purified by recrystallization.

Neutral potassium oxalate, K$_2$C$_2$O$_4$ + 2H$_2$O, and **Potassium cyanide,** KCN, are more powerful reducing agents than sodium carbonate, and in many cases are to be preferred. The cyanide is usually mixed with an equal amount of sodium carbonate.

Potassium nitrate, KNO$_3$, **Sodium nitrate,** NaNO$_3$, and **Potassium chlorate,** KClO$_3$, serve only as oxidizing agents.

Acid potassium sulphate, HKSO$_4$. It is employed in the fused state as a coarse powder, and must be kept in a bottle provided with a ground-glass stopper. It is used as a decomposing agent, and often expels volatile substances, which may be recognized by their odor or color.

Fused borio acid, B_4O_3. It is employed in the state of a coarse powder, especially for the detection of small quantities of copper in lead.

Fluorite, CaF_2. Must be deprived of water by ignition; must be perfectly free from boric acid, for which it may be tested as described par. 75. It is convenient to keep in a separate bottle a mixture of one part of finely-powdered fluorite, with four parts of acid potassium sulphate.

Cobalt nitrate, $Co(NO_3)_2 + 6H_2O$, in solution. It must be pure, free from alkali, iron and nickel. The solution should not be too concentrated — about one part of nitrate to ten of water; and, as only one or two drops are used at a time, it is convenient to keep it in a bottle provided with a long stopper or pipette for the purpose of dropping, Fig. 25, or, still better, in a small bulb, Fig. 26, blown from a thick, soft glass tube about 20 to 25 mm. in diameter. To fill the bulb it is gently heated and the tip placed in the solution.

Fig. 25.

Fig. 26.

After a drop has entered the bulb, it is converted into vapor by carefully heating, and the tip again placed in the liquid, which immediately flows into the bulb. It should not be more than half or two-thirds full. If now the bulb is inverted and held in the warm hand, the solution is forced out by the expansion of the air. It is used for the detection of earths and metallic oxides, which **give** characteristic colors on being moistened and heated with it.

Nickel oxalate. It must be perfectly free from iron and cobalt ; it is tested with borax, with which it ought to produce a pure brown glass.

Hydrochloric acid, HCl, is used for the decomposition of carbonates, for conversion of substances into chlorides, in testing flame-coloration, and, in connection with zinc, for the detection of some of the rare metals.

Nitric acid, HNO₃, is used in the separation of silver from gold.

Sulphuric acid, H₂SO₄. Pure concentrated, for various purposes, especially testing for flame-colors.

Glycerin, for detection of boric acid, as proposed by Iles, for decomposition of silicates to extract their alkalies, for spectroscopic work, and as a general laboratory reagent when mixed with two parts of HCl, as proposed by Leeds.

Copper oxide, CuO. It is best prepared by igniting the dried nitrate in a porcelain dish.

Silver chloride, AgCl. It is prepared by precipitating a solution of silver nitrate with hydrochloric acid, washing the precipitate, and making it into a thick paste with water, which is kept in a small glass-stoppered bottle. This reagent should not be used with platinum wire, since the silver fuses with the platinum to an alloy ; fine iron wire is in this case substituted for the platinum. For each experiment a fresh hook should be made. It is used for intensifying the flame-coloration.

Lead. It is easily obtained pure by decomposing a solution of the acetate by metallic zinc ; the precipitate is repeatedly washed with water and then dried between filter papers.

Iron. In the shape of fine wire; used for reductions in the wet way.

Tin. Usually in the shape of foil, which is cut into strips and rolled up tightly into small long pellets. If applied to a bead containing a metallic oxide, and heated in the reducing flame, it acts as a powerful reducing agent.

Magnesium wire, in short pieces, for the detection of phosphoric acid.

Zinc, granulated, to be used with hydrochloric acid for the detection of rare metals whose solutions are reduced by the nascent hydrogen, giving characteristic changes of color.

Silver, foil or coin, for testing sulphur compounds.

Gold, in small grains, 50 to 60 mgrs., to use in testing for nickel and cobalt.

Bone-ash. In the state of very fine powder, for cupellation.

Test Papers. Blue and red litmus paper for the detection of acids and bases, Brazil-wood paper for the detection of fluorine, and turmeric paper for the alkalies, boric, and molybdic acids.

6. If the analytical research is strictly confined to blowpipe operations, the above reagents are sufficient; but if, as is sometimes advantageously done, some simple operations of the humid method of chemical analysis are called to aid, the list must be somewhat extended. The most important of these reagents are: potassium hydrate, ammonium hydrate, ammonium chloride, ammonium carbonate, ammonium oxalate, ammonium sulphide, ammonium molybdate, potassium ferrocyanide, potassium ferricyanide, platinum dichloride, lead acetate, alcohol, and distilled water.

The manner of using the balance is extremely simple. Before the substance is placed on the pan, the place of the mark is observed on the scale. A known weight is then placed on the upper pan, and the shelf, B, moved down as far as necessary—so far that, with the consequent extension of the spiral, the pan d will again sink into the water, when a second reading is made. The difference in these figures gives the number of divisions on the scale over which the spiral has been made to pass by the weight. If it is found, for example, that with a weight of one gram the extension of the spiral is 122.2 divisions, while with some substance, as a piece of mineral, the extension is only 54.4, the absolute weight of the substance will be $\frac{54.4}{122.2} = 0.445$.

If, however, the absolute weight is not desired, only the specific, it is not necessary to express the absolute weight in grams. Three readings are made : first, with empty pan ; second, with substance placed on the upper pan ; and the third after placing the same substance on the pan under water. The difference between the first two gives the absolute weight, expressed in divisions of the scale, and the difference in the last two gives the weight of the displaced water. The quotient of these differences is therefore the specific weight. If the mark with empty pans stands at 64.2, and with the substance placed in the upper pan at 275.3, and with the same substance in lower pan at 220.8, then the absolute weight is $275.3 - 64.2 = 211.1$, and loss of weight in water is $275.3 - 220.8 = 54.5$. Specific weight will be $\frac{211.1}{54.5} = 3.85$. The second decimal is not always certain, but by proper arrangement of the spiral is found as reliable as the ordinary balance.

If the specific weight of a fluid is to be determined,

both pans are taken off, and in place of them a glass of
about 1 cc. in size is suspended by a fine platinum wire.
The loss of weight in water and in other liquids is shown
by the scale, as in the former case.

As shown in the drawing, the shelf, B, is attached to the
standard, A.   The movable standard, C, has the same
length as A, can be raised or lowered, and made fast at
any point.   C is drawn out, according to the length of
the spiral, so far that the mark with the pans empty stands
opposite one of the upper divisions of the scale, which,
to show the whole extension of the spiral, must be at least
600 mm. long.

Every spiral at first shows a little elasticity, which
grows less, and which during any one experiment amounts
to really nothing.

OF THE FLAME, AND GENERAL ROUTINE OF BLOWPIPE ANALYSIS.

THE flame of a candle consists of three distinct parts—the dark central zone or supply of unburnt gas surrounding the wick; the luminous zone or area of incomplete combustion; and the non-luminous zone or mantle of complete combustion. (Fig. 28.) In this outer zone the supply of oxygen is greatest, all the carbon is at once burned, and the flame becomes non-luminous. The effect of producing a complete combustion at once throughout the flame is seen in the Bunsen gas-lamp or burner. In this lamp (see Fig. 4) the gas issues from a small central burner, and, passing up the tube, draws air with it through the holes at the bottom of the tube; the mixture of air and gas can be lighted at the top of the tube, where it burns with a non-luminous flame. If the holes be closed, the gas alone burns with the ordinary bright flame. The blowpipe flame may also be divided into two distinct parts—the *oxidising* flame, where there is excess of oxygen, and the *reducing* flame, where there is excess of carbon—and these are distinguished by the same properties as the outer, and inner or luminous, zone of the candle-flame.

Fig. 28.

To produce the *oxidizing* flame (Fig. 29), **the stream** of gas should not be **too strong.** The jet of the blow- pipe is placed just within the flame, near the slit in **the** tube, so that a *strong* current of air is thoroughly **mixed** with the gas, which forms an inner long blue flame, *a b.* The hottest part of the flame is just before the **apex of**

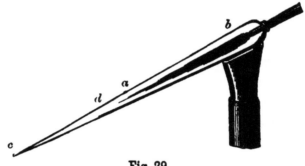

Fig. 29.

this blue cone, *d*, where the combustion of the **gases is** most complete. For fusion, substances are **exposed to** this part, but for oxidation are placed a little **beyond** the apex, exposed to the air.

The *reducing* flame (Fig. 30) is produced **by so placing**

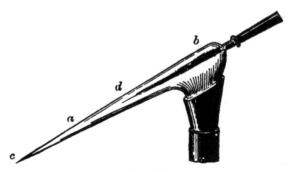

Fig. 30.

the blowpipe that the jet just reaches the flame a little above the slit, and a *gentle* current of air made to pass a little higher above the tube than in Fig. 29. The whole flame now appears as a long, luminous cone, *a b*, being partially charged with glowing carbon surrounded with a pale-blue mantle, which extends to *c*. The most active part of the flame lies between *a* and *d*, somewhat nearer *a*. Any reducible metallic oxide placed at this point will be deoxidized or reduced, on account of the tendency of the free carbon in the flame to combine with the oxygen. It is often found necessary to maintain the reducing flame for a considerable length of time, and that the substance should be completely surrounded by it, in order to obtain the desired results. On this account a little more practice is required for this flame than for the oxidizing.

The blast should be produced with the cheek-muscles alone, and not with the lungs. The trumpet-shaped mouth-piece is pressed against the lips, and breathing is effected through the nostrils. In this way a constant and regular blast may be produced, and, after a little practice, without any perceptible exertion or weariness.

7. On examining a substance before the blowpipe with a view to determine its nature, or to ascertain the presence or absence of certain matter, it is advisable to follow a systematic course, composed of a series of operations, and to attentively observe the changes which the body undergoes under the influence of the various agents which are brought to act upon it. The various operations to which the assay is submitted are so many questions, to which the phenomena we observe constitute so many answers; and from their appearance or non-appearance we are able to draw definite conclusions as to the nature of the substance under examination.

The following order, and the rules to be observed

in the execution of the various operations, are essentially the same as first pointed out and laid down by Berzelius.

(1.) Examination in a glass tube sealed at one end or a matrass. (Figs. 18 and 19.)

(2.) Examination in a straight or slightly-bent glass tube open at both ends. (Figs. 16 and 17.)

(3.) Examination on charcoal or aluminium foil by itself.

(4.) Examination with reagents—borax, salt of phosphorus, sodium carbonate, sodium thiosulphate, acid potassium sulphate, cobalt nitrate solution, and zinc with hydrochloric acid.

(5.) Examination in the platinum-pointed forceps, or on platinum wire by itself, for the determination of fusibility and flame-coloration.

Regarding the size of the assay, a piece the size of a mustard-seed will generally be found sufficient, larger pieces, without showing the reaction more distinctly, requiring more time and labor. In some cases, however, it is advantageous to employ a greater quantity, especially for reductions or for heating in a glass tube; for the larger the metallic globule, and the greater the amount of the sublimate produced, the more readily can its nature be ascertained. A portion of the original substance should be reserved for confirmatory tests or for examination in case of accident.

It is a good plan to place the lamp on a large piece of white paper with upturned edges, or on a bright tin plate, so that if a globule or portion of the specimen is dropped, it may be easily found.

EXAMINATION IN A GLASS TUBE CLOSED AT ONE END OR A MATRASS.

8. The substance is introduced into a small glass tube sealed at one end or into a small matrass, and heat applied by means of a gas- or spirit-lamp. The heat must at first be gentle, but may be gradually raised to redness if necessary. By this treatment we learn whether—

I. THE SUBSTANCE IS ENTIRELY OR PARTIALLY VOLATILE.

Among the phenomena to be observed, the following are deserving of particular attention :

9. (1.) **Water is given off**, which partly escapes and partly condenses in the colder portion of the tube. This points to the presence of a salt containing water of crystallization* [No. 46], to the presence of a hydrate, or to salts which contain water mechanically inclosed between the laminæ of the crystals [No. 41]; in this case the substance usually decrepitates. The drops of condensed water are to be examined with test-paper ; an alkaline reaction denotes the presence of ammonia, and an acid reaction the presence of some volatile acid, as sulphuric, nitric, hydrochloric, hydrofluoric, etc.

10. (2.) **Gas or vapor is given off.** Those of most usual occurrence are :

a. Oxygen, easily recognized by placing a small piece of coal upon the assay, which burns brilliantly on being heated ; points to the presence of a peroxide, nitrate,

* The numbers refer to the list of substances at the beginning of the book, page 16.

4

chlorate, bromate or iodate [No. 42]. If the quantity of the substance is very small, add a very little sodium chloride and sulphuric acid. On being warmed, in place of the oxygen, chlorine is set free, and is recognized by its smell and bleaching action on blue litmus paper.

b. Sulphur dioxide, easily recognized by its peculiar odor and action on blue litmus paper; indicates the presence of a sulphate or sulphite [No. 46].

c. Sulphuretted hydrogen, recognized by its peculiar odor; indicates the presence of sulphides containing water.

d. Nitrogen tetroxide, recognized by its deep orange-red color and acid reaction; indicates the presence of a nitrite or nitrate [No. 43].

e. Carbon monoxide, which burns with a blue flame; indicates oxalates or formates. In the latter case the substance blackens.

f. Carbon dioxide, recognized by causing a turbidity in a drop of lime-water suspended from the convex side of a watch-glass and exposed to the escaping gas; points to the presence of a carbonate or the oxalate of a reducible metallic oxide.

g. Cyanogen, recognized by its peculiar odor and by burning with a crimson flame; indicates the presence of a cyanogen compound.

h. Ammonia, recognized by its odor and alkaline reaction; indicates the presence of an ammoniacal salt or of an organic nitrogenous substance; in the latter case the mass usually blackens, and evolves at the same time either cyanogen or empyreumatic oils of offensive odor [No. 34].

i. Hydrofluoric acid changes the color of test-paper,

and also attacks the glass just above the assay and makes it dull.

j. Chlorine, indicated by greenish-yellow fumes and its odor.

k. Iodine is indicated by violent fumes and its peculiar odor, and, if the amount is not too small, will form a **steel-gray sublimate.**

l. Bromine, indicated by its orange-colored vapor and its pungent odor.

11. (3.) **A sublimate is formed.**

(*α.*) *WHITE SUBLIMATES* are formed by:

a. Ammonium salts. On removing the sublimate from the tube, placing it on a watch-glass, adding a drop of potassium hydrate and applying heat, ammonia is evolved [No. 37].

b. Mercury chlorides. The mercurous chloride sublimes without previous fusion; the mercuric chloride fuses first, then sublimes; the sublimate is yellow while hot, but becomes white on cooling [Nos. 49 and 50]. Mercuric oxide forms globules of mercury.

c. Antimony trioxide. It fuses first to a yellow liquid, then sublimes; the sublimate consists of lustrous needle-shaped crystals [No. 19].

d. Arsenic trioxide. The sublimate consists of octahedral crystals [No. 31].

e. Tellurium dioxide shows a reaction similar to that of antimony trioxide, but requires a much higher temperature; the sublimate is amorphous.

f. Osmium tetroxide forms a sublimate of white drops, with a pungent, disagreeable odor.

(β.) *BLACK OR GRAY SUBLIMATES* with metallic lustre, or metallic mirrors, are formed by:

a. Arsenic, and *arsenides* containing more than one

equivalent of arsenic to two of metal; also, some sul
arsenides [No. 74]; cutting the tube below the sublim
and exposing the mirror to gentle heat in the gas-fla1
the garlic odor of arsenic is perceived.

b. Mercury, amalgams, and some mercury salts;
sublimate consists of minute globules of mercury, whi
by friction with a piece of copper wire, readily unite
larger globules [No. 9].

c. Some alloys of *Cadmium.*

d. Tellurium, only at a very high temperature;
sublimate consists of small globules, which solidify
cooling.

(γ.) *COLORED SUBLIMATES* are formed by:

a. Sulphur, and *Sulphides* containing a large amoun1
sulphur; the sublimate is deep-yellow to brownish-
while hot, but pure sulphur-yellow when cold [No. 7

b. Antimony sulphides, alone or in combination w
other sulphides; the sublimate forms only at a very h
temperature, and is deposited at a short distance fr
the assay-piece; it is black while hot, reddish-bro
when cold [No. 71].

c. Arsenic sulphides and some compounds of meta
sulphides with arsenides; the sublimate is dark browni
red while hot, but reddish-yellow to red when cold [1
77].

d. Cinnabar. The sublimate is black, without lus1
and sometimes yields a red powder on being rubbed
if scratched with a knife [No. 78].

e. Selenium and some *selenides;* the sublimate appe
only at a high temperature, is of a reddish or black col
and yields a dark-red powder; at the open end of
tube the peculiar odor of selenium (resembling rot
horse-radish) is perceived [No. 84].

II. THE SUBSTANCE CHANGES WITHOUT VOLA-TILIZATION.

12. Many substances under this treatment suffer physical changes without being affected in their chemical constitution. The most important of these physical changes are .

1. **Change of color:**

(*a.*) From white to yellow, and white again on cooling; zinc oxide [No. 22].

(*b.*) From white to yellowish-brown, dirty pale yellow on cooling; tin oxide [No. 23].

(*c.*) From white to brownish-red, yellow when cold, and fusible at a red heat; lead oxide [No. 69].

(*d.*) From white to orange-yellow or reddish-brown, pale yellow when cold, and fusible at a bright red heat; bismuth oxide [No. 20].

(*e.*) From red to black, and red again on cooling; ferric oxide (*non-volatile*) [No. 24].

(*f.*) From red to black, red when cold; mercuric oxide (*volatile*).

2. **Carbonization:** organic substances.

3. **Fusion:** some alkaline salts.

4. **Decrepitation:** alkaline chlorides and very many minerals.

5. **Phosphorescence:** alkaline earths, earths, tin oxide, zinc oxide, and many minerals.

EXAMINATION IN A GLASS TUBE OPEN AT BOTH ENDS.

13. A fragment of the substance, sometimes in form of a powder, is introduced into the tube to a depth of 10 or 12 mm., the end to which it lies nearest slightly inclined, and heat applied. The air contained in the tube becomes heated; it rises, escapes from the upper end, and fresh air enters from below. In this manner a calcination is effected, and many substances which remained unchanged when heated in a matrass yield sublimates or gaseous products when subjected to this treatment, owing to the formation of volatile oxides.

By this means the presence of the following substances can be detected:

14. Sulphur. Sulphur dioxide is evolved, which is characterized by its peculiar odor and action on moistened blue litmus paper [No. 72].

15. Arsenic. If present in sufficient quantity it yields a white and very volatile sublimate of arsenic trioxide, consisting of minute octahedral crystals; by application of gentle heat it may be driven from one place to another [No. 74].

16. Antimony. White fumes of antimony trioxide are given out, which partly escape and partly condense in the upper part of the tube. The sublimate is a white powder, and may, if consisting of pure antimony trioxide, be volatilized by heat. In most cases, however, the oxidation proceeds farther, and antimony tetroxide, a non-volatile white powder, is formed [No. 1].

17. Bismuth. When not combined with sulphur, it is converted into oxide, which condenses at a short distance from the assay, and which by heat may be fused to brownish globules, which on cooling become pale yellow [No. 4].

18. Mercury and **Amalgams** yield sublimates of metallic mercury in small globules [No. 9].

19. Tellurium and **Tellurides.** Tellurium dioxide is produced, which condenses in the upper part of the tube to a white non-volatile powder; on application of heat it fuses to colorless globules, thus distinguishing it from antimony [No. 85].

20. Selenium and **Selenides** evolve a gaseous oxide of a peculiar odor, resembling that of rotten horse-radish [No. 84]; a sublimate of selenium, gray near the assay and red at a distance, is sometimes formed.

EXAMINATION ON CHARCOAL OR ALUMINIUM FOIL.

21. A small quantity of the substance is placed in a shallow cavity near to the edge of the coal, which is held slightly inclined, so that when the flame is directed upon it, the coating, if one is formed, is deposited on the coal. Its behavior in both flames should be observed. If the substance is in the form of a powder, or if it decrepitates so that it must be reduced to a powder, it may be moistened with water and then packed into the cavity. Or a small grain of borax may be first fused in the cavity, and the powder or fragment placed upon it while the borax is still in the melted state.

Should aluminium foil be used in place of charcoal, the substance is first heated on the bare foil, and afterward on the small piece of coal described on page 22. In this way the readily volatile metals, together with the difficultly ones, may often be determined. The coatings are thicker on the foil than on the coal because the support remains at a lower temperature and the vapors condense upon the vertical portion. The coatings are treated with the reducing flame and the peroxidizing flame. The latter is obtained by bringing a good oxidizing flame within one or two inches of the coating obtained.

The following phenomena should be considered:

22. (1.) **Fusibility.** The *easily-fusible non-metal-*

lic compounds are most of the alkaline salts and some of the salts of the alkaline earths, as barium and strontium hydrates, and, after continued heating, their carbonates and sulphates. Their residues, after ignition, have an alkaline reaction, turning moist turmeric paper brown. Some are volatile and cover the charcoal with a coating (see page 52 (1, 2, 3).

The *infusible without flame-coloration* are compounds of the earths and of the alkaline earths, which glow intensely when heated, and may be further tested with cobalt solution; also silica and many silicates.

The *infusible with change of color* are the oxides of zinc, tin, titanium, niobium, tantalum, tungsten, which take on a yellow color while hot.

The *easily fusible metals* are antimony, lead, bismuth, cadmium, zinc, tin, tellurium, thallium, and indium.

The *somewhat difficultly fusible* are gold, silver, and copper.

The *infusible* are iron, cobalt, nickel, molybdenum, platinum, iridium, osmium, palladium, rhodium, and tungsten.

(2.) **Deflagration.** Nitrates, chlorates, iodates, bromates.

(3.) **Decrepitation.** Sodium chloride, other haloid salts, substances containing water, and many minerals.

(4.) **Intumescence.** Substances containing water of crystallization, as borates, alums, etc.

(5.) **Odor.** It is usually given off as soon as the flame is directed upon the substance. Smell of sulphur dioxide indicates sulphur or sulphides; odor of garlic, arsenic; odor of rotten horse-radish, selenium.

(6.) **Color of flame** may be better seen when platinum wire or the platinum-pointed forceps are used for support and the substance placed in the Bunsen flame (see pars. 56–61 and Chap. IV.).

23. (7.) **Reduction to metal and formation of a coating.** Some metallic oxides are reduced to metal when heated on charcoal, some are wholly or partially volatilized, the vapor passing off or forming a coating upon the charcoal. Oftentimes a considerable amount of ash is formed upon the coal where the flame strikes it, which may be mistaken for a coating.

As already observed, most of the metallic oxides can be easily reduced by the reducing flame alone, others with difficulty, and some not at all. The latter, in a fine state of division, may be mixed with sodium carbonate or with a mixture of this with potassium cyanide, or potassium oxalate if it be used, and then treated in the reducing flame. The reduction is generally easily accomplished.

The above reagents do not prevent the formation of a coating.

(A.) REDUCED METAL WITHOUT COATING.

24. Gold, silver, and copper form malleable beads with metallic lustre. Molybdenum, tungsten, platinum, palladium, iridium, rhodium, iron, nickel, cobalt, give a gray infusible powder, which, in case of the last three substances, is magnetic.

To separate the reduced metal, the fused mass is cut out from the charcoal, ground with water in a mortar, and the lighter particles of coal are poured off with the water. The malleable metals remain in flattened, shining scales, and the brittle ones as metallic powder. Silver, gold, and copper may be distinguished by their colors.

The other metals are determined by further treatment
with borax and salt of phosphorus.

(B.) REDUCED METAL WITH COATING.

25. Antimony. It fuses readily and covers the charcoal with white oxide; the ring is not so far distant from
the assay-piece as in the case of arsenic; it may be driven
about by the oxidizing flame and made to disappear with
the reducing flame, which it colors a very pale green, but
is not so volatile as that of arsenic, and does not emit an
alliaceous odor. Metallic antimony, when fused on charcoal and heated to redness, remains a considerable time in
a state of ignition without the aid of the blowpipe, disen-
gaging, at the same time, a thick white smoke, which is
partly deposited on the charcoal around the metallic globule in white crystals of a pearly lustre [No. 1].

On aluminium foil the coating near to the assay is yellow, farther away pure white, and still farther off bluish-
white. Most minerals yield antimony on the bare plate.
The peroxidizing flame darkens the yellow color momentarily, whilst the reducing flame instantly blackens all
parts of the coating.

26. Bismuth. It fuses readily in both flames and covers the charcoal with oxide, which is dark orange-yellow
while hot and lemon-yellow when cold. The yellow
coating is usually surrounded by a yellowish-white ring,
consisting of bismuth carbonate. The coating is somewhat nearer the assay than that of antimony; it may be
driven away by both flames, but, unlike antimony and
lead, does not impart any color to the reducing flame
during the operation [No. 4].

On the aluminium foil little or no coating is obtained, but on the charcoal support a coating is produced

which is yellow nearest the test-piece, passing into orange, and this into brown. Upon the ledge also a yellow coating is formed. The oxidizing flame darkens the color of the yellow and orange portions temporarily (compare *Lead*), whilst the reducing flame blackens both.

27. Lead. It fuses easily and coats the charcoal in both flames with oxide, which is dark lemon-yellow while hot and sulphur-yellow when cold, with a border of bluish-white which consists of carbonate. The coating is found at the same distance from the assay as that of bismuth; it may be driven away by either flame; when played upon with the reducing flame it imparts to it an azure-blue color [No. 3].

On aluminium foil a coating is obtained only on using a charcoal support, as in the case of bismuth. The coating is coffee-brown, surrounding a pale yellow ring; on the ledge it is white. The yellow and white parts become brown in the oxidizing flame, retaining this color on cooling, whilst the brown color, in the case of bismuth, disappears again on cooling. In the reducing flame all parts become black.

28. Tin. It fuses readily; exposed to the oxidizing flame, it is converted into oxide, which may be blown away and thus be made to appear as a coating; it is always found closely surrounding the assay-piece, is slightly yellow and luminous while hot, white when cold, and non-volatile in both flames. Exposed to the reducing flame, the molten metal retains its bright metallic aspect [No. 7].

On aluminium foil a faint white coating is obtained by long heating on the charcoal support. The reaction on the support is the same as that on charcoal, just described.

29. Silver. When exposed for a long time to the action of the reducing flame it yields a slight dark-red coating of oxide [No. 8]. If the silver contains lead or antimony, a yellow or white coating appears before the red one; or if lead and antimony are present at the same time, the coating has a bright rose-color.

On aluminium plate a brown coating, shading off into a lighter rim having a reddish tinge, is obtained. Near the glowing edge of the charcoal is a narrow whitish strip with a faint pink tinge. The oxidizing flame darkens all parts, whilst the reducing flame produces a circle of white having the appearance of frosted silver. The rose-colored coating produced by silver in presence of antimony, coming out beautifully on aluminium plate, is, however, more characteristic.

30. Gold. On charcoal fuses, but gives no coating.

On aluminium plate a coating is produced after heating for some time with charcoal support. The gold is volatilized, and near the charcoal a yellow film of gilding is deposited on the plate; beyond this is a strip of violet, and dotted all over are little specks of gold carried away mechanically.

31. Thallium. It fuses easily and coats the charcoal with white oxide, which is driven away by slight warming; on contact with the flame this latter acquires a green coloration, and the oxide disappears. The fused bead, which also colors the flame green, remains fluid for a considerable time after the flame is removed, and sometimes deposits a brown coating in its neighborhood.

On aluminium plate a copious white coating is first produced, which, as the temperature of the plate increases, is followed by a brownish one. In the oxidizing flame the white part instantly turns reddish-brown.

This change takes place more quickly than with lead, and the color produced is very different. In the reducing flame all parts become black, and in the thickest portions little black beads can be seen with a lens.

32. Indium. It fuses readily, forming a coating very near the assay, which is dark yellow while hot and yellowish white when cold. It may be driven off with difficulty by the reducing flame, to which it gives a clear violet tint.

(C.) COATING WITHOUT REDUCED METAL.

33. Arsenic. It is volatilized without previous fusion; the charcoal is covered with a white coating, which is far distant from the assay-piece, and which is produced by both the oxidizing flame and reducing flame; the coating is very volatile, and is easily driven away by the blow-pipe flame, to which it imparts a light-blue color, emitting the peculiar alliaceous odor characteristic of arsenic [No. 2].

On aluminium foil without charcoal support a white coating is obtained, and a black stain is produced under the test-piece. On the charcoal support, when much arsenic is present, there is also a grayish-black coating, together with large black stains on the ledge. In the oxidizing flame the white portion is unchanged, but volatilizes rapidly as the plate gets hot. The gray and black portions are somewhat whitened and partly removed, but dark stains remain. The reducing flame volatilizes the coating rapidly, and the arsenic smell is very clearly perceptible.

34. Zinc. It fuses readily; exposed to the oxidizing flame it burns with an intensely luminous greenish-white flame, emitting at the same time a thick white smoke, which, partly condensing on the charcoal, rather near

the assay, covers it with oxide, yellow while
white when cold. The coating, when played u₁
the oxidizing flame, becomes luminous, but does
appear [No. 6].

On aluminium foil without charcoal support,
any coating is obtained. On a charcoal support
film is produced as soon as the metal begins to bur₁
directly gives place to the white oxide film. T₁
izing flame and reducing flame have no action
coating. Minerals containing zinc do not give t₁
coating.

35. Cadmium. It fuses readily, and exposed
oxidizing flame it burns with a dark-yellow flam
ting brown fumes of oxide, which cover the
around and near the assay. This coating is v₁
racteristic; it is, when cold, of a reddish-browr
in thin layers, orange-yellow; it is easily volatil
both flames without imparting a color to then
yond the coating a variegated border is sometim
[No. 5].

On aluminium foil a dark brown, almost blacl
obtained, which is not affected by either the o
flame or reducing flame. On the edges of the
support a little reddish-brown oxide is usually de₁

36. Selenium. It fuses very readily in both flan
disengagement of brown fumes; at a short distan
the assay a steel-gray coating of a feeble metalli₁
is deposited; played upon with the reducing
disappears with emission of a strong odor o
horse-radish, at the same time imparting to the
fine blue color [No. 84].

On aluminium foil with charcoal support a red
is formed, together with some brown and white fil₁

oxidizing flame whitens the red and brown parts, whilst the reducing flame gives to all parts a deep brown color.

37. Tellurium. It fuses very readily and coats the charcoal in both flames with tellurium dioxide; the coating is not very far distant from the assay; it is of a white color with a red or dark-yellow edge; played upon with the reducing flame it disappears, imparting to the flame a green tinge.

On aluminium foil, with and without the charcoal support, a strong coating is formed close to the assay, which is brown in thin films. Where the deposit is thickest a white layer of tellurium dioxide forms on short exposure to the oxidizing flame. The reducing flame turns all parts black; on longer blowing the coating disappears, the flame becoming tinged with green.

38. Molybdenum. The metal, a grayish infusible powder, oxidizes in the oxidizing flame and gives a partly crystalline coating, which is yellow whilst hot and white when cold. By momentary exposure to the flame the coating becomes of a beautiful dark-blue color (molybdic molybdate); by longer heating it becomes dark copper-red, with metallic lustre (molybdenum dioxide).

On aluminium plate the coating (best obtained from molybdenite or ammonium molybdate) is produced without charcoal and is light-yellow with white film. The oxidizing flame darkens the color somewhat, whilst the reducing flame by momentary contact produces a beautiful blue coloration.

Besides the above-named elements, there are other substances that yield white coatings which may, with few exceptions, be driven away when played upon by the oxidizing flame, and which bear some resemblance to those described. The most important bodies of this kind are the following:

(1.) The sulphides of the alkalies, of lead, bismuth, antimony, zinc (coating non-volatile), tin (coating non-volatile), and the chlorine, iodine, and bromine compounds of ammonium, mercury, and antimony: they coat the charcoal without previously fusing or sinking into the support.

(2.) The compounds of the alkalies with chlorine, bromine, iodine, and sulphuric acid; they fuse and sink into the charcoal before they evaporate.

(3.) The chlorine, bromine, and iodine compounds of lead, tin, bismuth, zinc, and cadmium, which fuse but do not sink into the charcoal before they coat it.

EXAMINATION WITH BORAX AND SALT OF PHOSPHORUS.

39. The examination of the assay with borax and salt of phosphorus is eminently adapted to detect the presence of metallic oxides, a great number of them possessing the property of being at a high temperature dissolved by these fluxes with a characteristic color. Unoxidized metals and metallic sulphides, arsenides, etc., differ in this respect very materially from the pure oxides; hence it is necessary before performing the experiment to convert all such substances into oxides. This is effected by calcination, or roasting on charcoal or in an open glass tube. The finely-powdered assay is placed on charcoal and alternately treated with the oxidizing flame and reducing flame, and this process is repeated until the substance no longer emits, while in the incandescent state, the odor of sulphur or arsenic. The heat must never be raised so high as to cause fusion, and between every two succeeding calcinations the assay should be taken from the charcoal and freshly powdered.

The experiment with borax is generally made on platinum wire, where the color of the bead is more readily observed; charcoal is used only in such cases where the substance under examination contains metallic oxides which are easily reduced to metal which attacks platinum. It is not sufficient to observe the color of the bead after cooling, but all changes of color which take place during the action of the flame, and through all the various stages of cooling, should be carefully noticed.

40. Flaming. In many cases when a transparent bead is intermittently heated in the flame, or is repeatedly taken out of the flame, peculiar effects are obtained. This operation has received the name of "flaming." Clear beads frequently become opaque, milk-white, or even colored. This depends on the fact that certain compounds which dissolve at a high temperature separate out on being heated to a somewhat lower temperature, appearing as peculiar crystals, which are sufficiently well formed in most cases to be visible under the microscope when the bead has been flattened whilst hot, or when it has been dissolved in dilute acid so as to isolate the crystals.

41. The behavior of the metallic oxides with borax and salt of phosphorus is shown in the following tables. They are arranged according to the color yielded by the hot bead when acted on by the oxidizing flame, and the reactions of the oxidizing and reducing flames are given in the same line. It may be here remarked, salt of phosphorus beads are often more beautiful than those of borax, and are occasionally different in color.

The behavior of the metallic oxides to these reagents is also given in the third and fourth columns of the table at the end of Chapter IV., where the metals are arranged in alphabetical order.

5*

BEHAVIOR OF METALLIC OXIDES WITH BORAX.

Contractions: fl. = by flaming; c. bl. = continued blowing; s. b. = saturated bead.

In the Oxidizing Flame.		In the Reducing Flame.		Indication.
Hot.	Cold.	Hot.	Cold.	
colorless	colorless	colorless	colorless	Silica
"	"	"	"	Alumina
"	"	"	"	Tin oxide
"	colorless; fl., opaque	gray; c. bl., colorless	gray; c. bl., colorless	Silver oxide
"	"	"	"	Tellurium dioxide
"	"	colorless	colorless; fl., opaque	Baryta
"	"	"	"	Strontia
"	"	"	"	Lime
"	"	"	"	Magnesia
"	"	"	"	Glucina
"	"	"	"	Yttria
"	"	"	"	Zirconia
"	"	"	"	Thoria
"	"	"	"	Lanthanum oxide
"	"	"	"	Tantalum pentoxide
"	colorless	s. b., rose-red	s. b., rose-red	Didymium oxide
"	colorless; fl., opaque	colorless; s. b., gray	colorless; s. b., gray	Niobium pentoxide
colorless; s. b., yellow	"	yellow to brown	yellow to brown; s. b. fl., blue	Titanium dioxide

BEHAVIOR WITH BORAX—*Continued.*

In the Oxidizing Flame.		In the Reducing Flame.		Indication.
Hot.	Cold.	Hot.	Cold.	
colorless; s. b., yellow	colorless; s. b., enamel-white	yellow	yellowish-brown	Tungsten trioxide
yellowish	"	gray; c. bl., colorless	gray; c. bl., colorless	Zinc oxide
"	colorless	"	"	Cadmium oxide
"	colorless	"	"	Antimony oxide
yellow	colorless; fl., opaque	"	"	Lead oxide
"	colorless; s. b., yellow and opalescent	"	"	Bismuth oxide
"	greenish-yellow	brownish	emerald-green	Vanadium pentoxide
yellow to red	colorless to yellow	green	bottle-green	Iron oxide
"	colorless to yellow; fl., idle	"	bottle-green; s. b., fl., black	Uranium oxide
"	"	colorless	colorless; s. b., enamel-white	Cerium oxide
"	colorless to opalescent	brown	opaque-brown *	Molybdenum trioxide
"	grass-green	green	emerald-green	Chromium oxide
violet	reddish-brown	gray; c. bl., colorless	gray; c. bl., colorless	Nickel oxide
violet: s. b., black	reddish-violet; s.b., black	colorless	colorless	Manganese oxide
blue	blue	blue	blue	Cobalt oxide
green	bluish-green	colorless	brown; c. bl., red	Copper oxide

* In a strong reducing flame black particles of molybdenum oxide appear in the yellow bead.

BEHAVIOR OF METALLIC OXIDES WITH SALT OF PHOSPHORUS.

Contractions: fl. = by flaming; c. bl. = continued blowing; s. b. = saturated bead.

In the Oxidizing Flame.		In the Reducing Flame.		Indication.
Hot.	*Cold.*	*Hot.*	*Cold.*	
silica skeleton	silica skeleton	silica skeleton	silica skeleton	Silica
colorless	colorless	colorless	colorless	Alumina
"	"	"	"	Tin oxide
"	colorless; fl, opaque	"	colorless; fl., opaque	Baryta
"	"	"	"	Strontia
"	"	"	"	Lime
"	"	"	"	Magnesia
"	"	"	"	Glucina
"	"	"	"	Yttria
"	"	"	"	Zirconia
"	"	"	"	Thoria
"	colorless	"	"	Lanthanum oxide
"	colorless	"	c. bl., violet	Didymium oxide
"	colorless; fl., opaque	gray; c. bl., colorless	gray; c. bl., colorless	Tellurium dioxide
colorless; s. b., yellow	colorless; s. b., milk-white	"	"	Zinc oxide
"	"	"	"	Cadmium oxide
"	"	"	"	Lead oxide
"	"	"	"	Antimony oxide

BEHAVIOR WITH SALT OF PHOSPHORUS—*Continued.*

| In the Oxidizing Flame | | In the Reducing Flame | | Indication. |
Hot.	Cold.	Hot.	Cold.	
colorless; s. b., yellow	colorless; s. b., milk-white	gray; c. bl., colorless	gray; c. bl., colorless	Bismuth oxide
"	colorless	colorless	colorless	Tantalum oxide
"	"	dirty-green	blue*	Tungsten trioxide
"	"	yellow	violet*	Titanium dioxide
"	"	blue or brown	blue or brown*	Niobium pentoxide
yellow	yellow. s. b., opalescent	gray; c. bl., colorless	gray; c. bl., colorless	Silver oxide
"	yellowish-green	dirty-green	beautiful regn.	Uranium oxide
dark-yellow	light-yellow	brownish	grn.	Vanadium pentoxide
yellow to red green	colorless to yellow or brown	yellow to red greenish	colorless to red-dish	Iron oxide
"	colorless	colorless	colorless	Cerium oxide
reddish to brownish-red	yellow to reddish-yellow	reddish	yellow	Nickel oxide
reddish dirty-green	emerald-green	"	green	Chromium oxide
violet	violet	colorless	colorless	Manganese oxide
blue	blue	blue	blue	Cobalt oxide
green	"	dark-green	brownish-red (turbid)	Copper oxide
"	faint yellowish-green, almost colorless	dirty-green	pure green	Molybdenum trioxide

* Blood-red on addition of iron.

EXAMINATION WITH SODIUM CARBONATE.

42. The examination with soda is usually perform
charcoal in the reducing flame, and as a general rul
flux is added successively in small portions.   It is :
times better to form the pulverized assay into a past
moistened soda before placing it upon the coal.   1
particularly necessary when the assay is to be teste
its fusibility with soda, since a great many mineral:
behave very differently with different quantities of th

43. Instead of sodium carbonate, the *neutral pota*
oxalate or *potassium cyanide* may be advantageousl
for all experiments of reduction, since these reagents
cise a more powerful reducing action than sodium
nate.   They are for this reason frequently employed
the presence of such metallic oxides is suspected
conversion into metals requires high temperatures ar
aid of a very efficient deoxidizing agent.

44. In subjecting a body to the treatment of sod
have to direct our attention to two points.

Some substances unite with soda to fusible compo
others form infusible compounds, and others again a
acted upon at all; in the last case the soda is absorb
the charcoal and the assay is left unchanged.   With
form fusible compounds with effervescence:

45. Silicic acid fuses to a transparent glassy bead w
after cooling, remains transparent if the soda has no
added in too great excess [No. 51].

Titanium dioxide fuses to a transparent glassy
which is dark-yellow while hot; on cooling be
opaque and crystalline [No. 62].

Tungsten trioxide and Molybdenum trioxide, aft
carbon dioxide is driven off, are absorbed by the
coal [No. 29 and No. 30].

Tantalum pentoxide, vanadium pentoxide, and niobium pentoxide also yield fusible compounds and sink into the charcoal.

Lime, magnesia, alumina, zirconia, thoria, yttria, and glucina, as well as cerium and uranium oxides, are not attacked; they remain unchanged, whilst the soda sinks into the charcoal.

The salts of barium and strontium form with soda fusible compounds which are absorbed by the charcoal [No. 54 and 55].

Sodium carbonate is also used for the detection of:

Sulphur, selenium, and *tellurium compounds*, which give with it a fused mass, yielding a black, brown, or yellow stain when laid on a piece of silver and moistened with water.

Manganese and *chromium*, with the soda alone, or better, with addition of sodium nitrate, yield colored masses; the former a green mass of manganate and the latter a yellow mass of chromate.

46. The second point to be observed is the elimination of metallic matter. Of the metallic oxides, when treated with soda on charcoal in reducing flame, are reduced: the oxides of the noble metals and the oxides of arsenic, antimony, bismuth, indium, cadmium, copper, cobalt, iron, lead, mercury, nickel, tin, zinc, molybdenum, tungsten, and tellurium. Of these, arsenic and mercury vaporize so rapidly that frequently not even a coating is left on the charcoal. Antimony, bismuth, cadmium, lead, zinc, and

The fused mass of soda and metal, and the portion of the charcoal immediately below and around the assay, are placed in a small mortar, rubbed to powder, the powder mixed with a little water and stirred up. The heavy metallic particles settle to the bottom, part of the soda dissolves, and the charcoal powder remains suspended in the water. The liquid is carefully poured off and the residue treated repeatedly in the same manner until all foreign matter is removed. The metal remains behind as a dark heavy powder, or when the metal is ductile and easily fusible, in the shape of small flattened scales of metallic lustre. These may be examined with the magnifying-glass, and also with the magnet. If the substance under examination contains several metallic oxides, the metallic mass obtained is usually an alloy, in which the several metals may be recognized by processes to be described hereafter. It is only in some exceptional cases that separate metallic globules are obtained, for example, in substances containing iron and copper.

For a more detailed account of the behavior of the various metallic oxides under this treatment, see the second column of Tables I. and II., pages 189–215.

A list of the oxidized minerals arranged according to their fusibility and behavior with sodium carbonate may be found under its appropriate head.

EXAMINATION WITH SODIUM THIOSULPHATE.

47. All the metals precipitated by sulphuretted hydrogen in the wet way, yield the sulphide reaction in the dry way when the substance is heated with powdered sodium thiosulphate. The reagent may be applied to a borax bead, in which the substance is already dissolved, the bead being then heated in the reducing flame. This method has, however, the disadvantages that easily-volatile substances, such as arsenic and mercury compounds, afford no reaction, and that the color imparted to the bead by the sulphide formed may easily be mistaken. Hence it is better to heat the powdered substance with the reagent in a glass tube closed at one end. After the decomposition of the thiosulphate, which is easily recognized by the odor of sulphuretted hydrogen produced, the color of the fused mass, due to the sulphide formed, is very readily seen.

In many cases the reaction is accelerated by the addition of a small quantity of oxalic acid.

As the thiosulphate contains a considerable amount of water of crystallization, the greater part of this should be previously expelled, or the glass should be held horizontally to prevent cracking, and have its mouth stopped with a little cotton-wool on first heating.

The sulphide reactions of the metals are given in the following table, together with the borax reactions. The two methods supplement each other exceedingly well:

Metallic oxide.	Reaction with $Na_2S_2O_3$.	Reaction with borax on platinum wire (the bead cold).	
		Oxidizing flame.	Reducing flame.
Antimony oxide......	red	colorless	gray to colorless
Arsenic "	yellow	....	....
Bismuth "	black	colorless	gray to colorless
Cadmium "	yellow	"	"
Chromium "	green	grass-green	emerald-green
Cobalt "	black	blue	blue
Copper "	"	bluish-green	brown
Gold "	"	reduced with	out dissolving
Iron "	"	yellow	bottle-green
Lead "	"	colorless	gray to colorless
Manganese "	light-green	reddish-violet	colorless
Mercury "	black	....	....
Molybdenum "	brown	colorless	brown

6

Metallic oxide.	Reaction with $Na_2S_2O_3$.	Reaction with borax on platinum wire (the bead cold).	
		Oxidizing flame.	Reducing flame.
Nickel oxide.........	black	reddish-brown	gray to colorless
Platinum "	"	reduced with	out dissolving
Silver "	"	colorless	gray to colorless
Thallium "	"	"	colorless
Tin "	brown	"	"
Uranium "	black	yellow	bottle-green
Zinc "	white	colorless	gray to colorless

EXAMINATION WITH ACID POTASSIUM SULPHATE or CONCENTRATED SULPHURIC ACID.

48. To determine the presence of volatile acids a small quantity of the substance is heated with acid potassium sulphate or with concentrated sulphuric acid (in the latter case, however, not to the boiling-point of the acid), and the following appearances are looked for:

1. **A colored gas is evolved.**

a. Nitrogen tetroxide fumes, known by their reddish-brown color and characteristic odor; evolved from nitrates and nitrites. With nitrates the reaction is promoted by the addition of metallic copper.

b. Chlorine tetroxide; yellowish-green, odor of chlorine, bleaching litmus paper, and explosive. The tetroxide is produced from chlorates by this treatment.*

c. Iodine, from iodides, is known by its violet vapors, which color starched paper blue. Iodates * give this reaction after the addition of ferrous sulphate.

d. Bromine; reddish-brown vapor, with pungent, unpleasant odor, and turning starch-paste yellow; yielded

* The chlorates, iodates, and bromates detonate when heated on charcoal.

by bromides and bromates. The color of the vapor is best seen on looking down the tube.

2. A colorless odorous gas is evolved.

49. *a. Sulphur dioxide*, from sulphites and polythionates, is easily known by its odor.

b. Hydrochloric acid, from chlorides, known by its odor and by the cloud of ammonium chloride which is formed when a glass rod moistened with ammonia solution is held near to the tube.

c. Hydrofluoric acid, from fluorides, has a very pungent odor and strongly corrodes glass.

d. Sulphuretted hydrogen, from sulphides, blackens paper moistened with lead acetate.

e. Cyanic acid, from cyanates, has a characteristic pungent odor; it brings tears into the eyes, and renders lime-water turbid.

f. Acetic acid, from acetates, is known by its pungent odor, and also by yielding fragrant acetic ether on heating with sulphuric acid and alcohol.

3. A colorless odorless gas is evolved.

50. *a. Carbon dioxide* is expelled from the carbonates with effervescence; it renders lime-water turbid.

b. Carbon monoxide, which burns with a bluish flame, may arise from oxalates, formates, cyanides, ferrocyanides, ferricyanides.

c. Chromic acid evolves oxygen, and the liquid turns brown or green.

d. Organic acids, recognized by the blackening due to the separation of carbon.

The acids which cannot be detected by the above methods, though easily detected in other ways, are: sulphuric, phosphoric, arsenic, boric, silicic, tungstic, molybdic, and titanic. With respect to the three last named, see par. 51.

BEHAVIOR OF METALLIC OXIDES WITH SALT OF PHOSPHORUS.

Contractions: fl. = by flaming; c. bl. = continued blowing; s. b. = saturated bead.

Indication	In the Oxidizing Flame Hot.	In the Oxidizing Flame Cold.	In the Reducing Flame Hot.	In the Reducing Flame Cold.
Silica	silica skeleton	silica skeleton	silica skeleton	silica skeleton
Alumina	colorless	colorless	colorless	colorless
Tin oxide	"	"	"	"
Baryta	"	colorless; fl, opaque	"	colorless; fl, opaque
Strontia	"	"	"	"
Lime	"	"	"	"
Magnesia	"	"	"	"
Yttria	"	"	"	"
Thia	"	"	"	"
Zirconia	"	"	"	"
Thoria	"	"	"	"
Lanthanum oxide	"	"	"	"
Didymium oxide	"	colorless	"	c. bl., violet
Tellurium dioxide	"	colorless; fl., opaque	gray; c. bl., colorless	gray; c. bl., colorless
Zinc oxide	"	colorless; s. b., milk-white	"	"
Cadmium oxide	colorless; s. b., yellow	"	"	"
Lead oxide	"	"	"	"
Antimony oxide	"	"	"	"

1 the platinum-pointed forceps, and treat-
xidizing flame.   Other substances must be
: powder placed on charcoal; moistened
)f cobalt solution, and treated as above.
1 only be distinguished after cooling.   A
f more or less purity, but rather dull, indi-
ence of alumina [No. 32], and a pale-red-
:sh-color) that of magnesia [No. 56].   It
;, be borne in mind that the alkaline and
licates, when heated with cobalt solution
ure above their fusing point, also assume
owing to the formation of cobalt silicate.
alumina, therefore, the heat must not be
as to cause fusion of the assay.   In testing
his precaution is not necessary; on the con-
r will appear the brighter and the more dis-
er the temperature to which the assay was
: alumina and magnesia reactions are pre-
presence of colored metallic oxides, which
luce a gray or black mass, unless present in
antity.
the oxides of the heavy metals, those of
:sume characteristic colors with solution of
reaction is best seen when the assay, alone
, soda, is exposed to the reducing flame on
ie ring of oxide which is deposited around
en moistened with solution of cobalt and
he oxidizing flame.   Zinc oxide takes a
-green and tin oxide a bluish-green color
lo. 23].
the compounds above mentioned, there
rs which, when exposed to the action of
1 and heat, experience a change of color.

E

These bodies are either of very rare occurrence, or the change produced in them is not sufficient to be of much importance. In fact, only a few colorations are of much use in the determination of substances—those of alumina, magnesia, zinc, and tin.

The following table gives the more definite colorations:

Blue: *Alumina;* deep color, infusible.

           Silica and *silicates;* faint color; with much solution of cobalt, black. Fine splinters fuse to a reddish-blue bead.

           Phosphates, silicates, and **borates** of the alkalies give a blue glass.

Violet: *Zirconia;* dirty-violet.

           Magnesium arsenate and *phosphate* fuse and become violet-red.

Flesh-red: *Magnesia;* pale flesh-red or pink.

           Tantalum pentoxide; hot, light gray; *cold,* flesh-red.

Brown: *Baryta; hot,* reddish-brown or brick-red; *cold,* colorless.

Green: *Zinc oxide*
           Titanium dioxide } ; yellowish-green.

           Tin oxide; bluish-green.

           Antimony oxide; dirty-green.

Gray: *Strontia;* dark-gray to black.

           Lime; gray.

           Glucina; bluish-gray.

           Niobium pentoxide; brownish-gray.

chloric acid, or sometimes with sulphuric, as in the case
of borates and **phosphates**.

The substance is exposed to the action of the inner
cone of the blowpipe flame, or what is much more convenient, the non-luminous flame of a Bunsen burner provided with a chimney. The colors are best seen against
a dark background and when there is not too much light.

If two or more elements which give color are present a
mixed color may be produced, or one element may over-
power the other; for example, sodium with potassium.
In this case the indigo prism, colored glasses, or still
better, the spectroscope, must be made use of. (See
pages 119, 120, 162.)

The colored flames given by the elements in a pure
state may be arranged as follows:

57. Yellow. *Sodium* and its salts cause an enlargement
of the outer flame and impart at the same time an intense
reddish-yellow color [No. 35]. The presence of other substances which also possess the property of coloring the
flame, but not in so high a degree, does not prevent the
reaction. Silicates containing sodium exhibit the same
phenomenon to a smaller or greater extent, according to
their degree of fusibility and the amount of sodium which
they contain [No. 59]. With many sodium salts which
do not exhibit the reaction very distinctly it can be produced by mixing the salt with some silver chloride to a
paste see par. 5), fastening it to the hook of a thin *iron*
wire, and then exposing it to the action of the inner
flame.

58. Violet. *Potassium* and most of its salts, with the
exception of borate and phosphate, impart to the outer
flame a distinct violet color [No. 38]. Also the salts of
rubidium and *cæsium* and the compounds of *indium,* but

these are rare as compared with the potassium. The presence of a sodium salt prevents the potassium reaction, in which case the flame is viewed through blue cobalt glass or a solution of indigo. Lithium also destroys the potassium flame, unless in very minute quantities. Potassium silicates must be free from sodium and lithium and easily fusible, at least on the edges.

59. Red. *Lithium* and its salts impart to the outer flame a fine *carmine-red* color [No. 60]; lithium chloride shows the reaction better than any other salt. The presence of a potassium salt does not prevent the reaction; the presence of even a small quantity of a sodium salt changes the color to yellowish-red, and a larger quantity prevents the reaction entirely.

Strontium chloride and some other strontium salts, for example, the carbonate and the sulphate, color the outer flame, immediately or after a while, *scarlet-red* [No. 54]. The presence of considerable barium prevents the reaction. Strontium carbonate and sulphate show the reaction remarkably well when mixed with silver chloride and heated on iron wire (see par. 5).

Calcium chloride, calcareous spar, many compact limestones, and fluorite, color the outer flame, immediately or after a while, *yellowish-red;* the color is not so intense as that produced by strontium. Gypsum and anhydrite impart at first a pale yellow, afterward a red color of little intensity [No. 52]. Fluorite gives at first a yellowish flame, but afterward an intense yellowish-red.

60. Green. *Barium* and its salts, especially after moistening with hydrochloric acid, color the outer flame *yellowish-green.* Silicates do not show the reaction. The presence of lime does not prevent the reaction [No. 55].

Copper oxide and some copper salts, as the carbonate and nitrate, impart to the outer flame a fine *emerald-green* color; after moistening with hydrochloric acid, blue. Compounds of iodine and copper and some silicates containing copper, as dioptase and chrysocolla, act in the same manner [No. 70]. *Thallium* and its salts color the flame grass-green.

Phosphoric acid, phosphates, and minerals containing phosphoric acid, especially if moistened with sulphuric acid, impart to the outer flame a bluish-green color, which is only seen for an instant [No. 34].

Boric acid colors the outer flame yellowish-green (greenfinch color); if a small quantity of sodium is present the color is mixed with yellow. Minerals containing boric acid should be pulverized and moistened with sulphuric acid.

Molybdenum oxides and molybdenite color the outer flame *yellowish-green*, in which the yellow is stronger than with barium [No. 30].

Tellurium dioxide fuses, emits white fumes, and color the outer flame green.

Calcium gives a *whitish-green* color, intensely bright.

Ammonium salts often show a *dark-green* flame, which, however, is very weak.

Nitric acid, bronze-green, quickly disappearing.

61. Blue. *Arsenic*, arsenic trioxide, and some arsenides, for example, smaltite and niccolite [No. 79], when heated on charcoal, impart a light-blue color to the outer flame. Some arsenates, for example, scorodite and erythrite, exhibit the same phenomenon in the forceps.

Antimony, fused on charcoal in reducing flame, is surrounded by a feeble greenish-blue flame [No. 1].

Lead, fused on charcoal in reducing flame, is surrounded by an azure-blue flame. Many salts of lead, heated in

\ the forceps or on platinum wire, impart an intense azure-blue color to the outer flame [No. 3].

Copper chloride colors the outer flame intensely *azure-blue;* after a while the color becomes green, owing to the formation of copper oxide [No. 47].

Copper bromide colors the outer flame greenish-blue; after a while the color changes to green.

Selenium, fused on charcoal in reducing flame, vaporizes with corn-flower-blue flame, with odor of rotten horse-radish.

In.lium colors both flames indigo-blue.

CHAPTER III.

SPECIAL REACTIONS FOR THE DETECTION OF CERTAIN SUBSTANCES WHEN IN COMBINATION WITH OTHERS.

62. THE preceding chapter and accompanying tables show the changes which many of the simple chemical compounds undergo when heated or when treated with the usual blowpipe reagents. The reactions are sufficiently characteristic to distinguish the various compounds from each other, so that when any one of the above-named substances in a pure state is under examination, there is no difficulty in determining its nature. This, however, is not of frequent occurrence, and in the majority of cases the body to be tested will be of a more complex nature. The results of the experiments will vary accordingly. For instance, a cobalt ore, containing iron, will not impart to the bead of borax or salt of phosphorus in the oxidizing flame a blue color, but a green one, resulting from the mixture of the blue of cobalt and the yellow of iron; lead, when accompanied by antimony, deposits a dark-yellow coating on charcoal resembling that of bismuth, etc. In such cases we may often, by attentively observing all the phenomena which present themselves, and by carefully comparing the results obtained by the various experiments, detect many, if not all, of the components of the substance under examination. Sometimes we attain this end more readily

by varying the order, or by introducing auxiliary agents into the series of experiments; and in other cases again it is only to be arrived at by subjecting the assay to treatments different from those mentioned in the preceding pages.

This chapter contains the principal reactions for the detection of substances which require the application of peculiar agents, and the methods for ascertaining the presence of certain bodies when in combination with others. The substances are arranged alphabetically.

63. Ammonia. Small quantities of ammonia are best detected by mixing the powdered assay [No. 37] with some sodium carbonate or potassium hydrate, introducing the mixture into a glass tube sealed at one end, and applying heat. The escaping gas is characterized by its odor and by its action on reddened litmus paper. White clouds are formed if a glass rod moistened with hydrochloric acid is held before the end of the open tube. From the appearance of this reaction we are, however, not authorized to infer the pre-existence of ammonia in the assay, since from organic matter containing nitrogen, when subjected to this treatment, ammonia is evolved as a product of decomposition.

Antimony. The reactions of antimony and its compounds, see pars. 11, 16, 25, 47, 55, 61, 153, Table II., 1.

64. Haamel's method for distinguishing antimony and other volatile metals, consists in moistening the metal or its oxide coating on coal with hydriodic acid, this reagent being made by passing sulphuretted hydrogen through water containing iodine until the solution becomes clear. On being heated the antimony coating thus formed is intense red, the cadmium white, the lead yellowish-green, and the bismuth brownish-red. In this way one metal

7

may be detected in the presence of **another, the color of** the iodides formed being so different and so intense.

Another method for the detection of **antimony in pres.** ence of *lead* or *bismuth* may be used.   The metallic com· pound [No. 10 or No. 82] is treated with fused boric acid on charcoal, the flame being so directed that the glass is always kept covered with the blue cone, the metallic glob. ule being on the side ; by this means the metals become oxidized, lead and bismuth oxides are absorbed by the boric acid, and the antimony oxide will form a ring on the charcoal, provided the temperature is not raised too high.

65. When combined with metals from which it is not easily separated, for example, *copper*, the evaporation of the antimony takes place so slowly that no distinct coat. ing is produced.   In this case the assay [No. 83] is treated with salt of phosphorus on charcoal in the oxidizing flame until the antimony, or at least part of it, has become ex. idized and entered into the flux.   The glass is now re. moved from the metallic globule and treated on another place of the charcoal with metallic tin in the reducing flame ; the presence of antimony will cause the glass to turn gray or black on cooling (Table II., 1).   Bismuth behaving under these circumstances in precisely the same manner, the presence of this metal makes the reaction not decisive for antimony.   The wet method has then to be resorted to.   (See par. 74.)

66. When the *antimony oxides* are accompanied by metallic oxides which, when reduced on charcoal, fuse with the metallic antimony to an alloy, as is the case with the *tin* and *copper* oxides, the latter cannot be recognized by a simple reduction.   The oxides have to be treated with a mixture of soda and borax on charcoal in the reducing

flame. The little metallic globules are separated from the flux and fused with from three to five times their own volume of pure lead and some fused boric acid in the reducing flame, care being taken to play with the flame only on the glass. Antimony oxide is volatilized, depositing the characteristic ring, while the oxides of the other metals are absorbed by the boric acid.

67. The *antimony sulphides*, when heated in the open glass tube, show the reaction mentioned in par. 16. When accompanied by *lead sulphide* [No. 18], only a small part of the antimony is converted into oxide, which sublimes; the remainder is changed into a white powder consisting of a mixture of antimony tetroxide, lead sulphate, and lead antimonate. For the detection of antimony proceed as in par. 74. When a compound containing *lead* or *bismuth sulphide*, besides *antimony sulphide*, is heated on charcoal in the reducing flame, a coating is deposited consisting of antimony tetroxide mixed with lead or bismuth sulphate, and, nearer to the assay, a yellow one of lead or bismuth oxide. In such a case the presence of antimony may be ascertained according to par. 101.

68. To detect a small amount of *antimony sulphide* in *arsenic sulphide*, Plattner recommends the following method: the assay [No. 17] is introduced into a glass tube sealed at one end and gently heated; the arsenic sulphide is volatilized and the greater part of the antimony sulphide remains as a black powder in the lower end of the tube; this end is cut off and the black substance taken out and transferred to a tube open at both ends. By applying heat the characteristic antimony reaction will appear.

Arsenic. The reactions of arsenic and its compounds, see pars. 11, 15, 22, 33, 47, 154, Table II., 2.

69. All metallic arsenides yield, when heat
open glass tube, a sublimate of arsenic trioxide
and most of them evolve a garlic odor (par.
heated on charcoal in the reducing flame [No. 7
recommends making the substance into a paste
coal powder and a dilute solution of shellac i
If this is made into small pencils and burnt, tl
given off. Some metals, for example, nickel ai
have a great affinity for arsenic, so that when on
quantity of the latter is present, the character
is not observable; in such cases it is sometimes
when the metallic compound is fused on cha
some pure lead in the oxidizing flame.

70. The *arsenic sulphides*, heated in the o
tube, evolve sulphur dioxide and yield a sub
arsenic trioxide. To show in a very decisiv
the presence of arsenic in any of its combinat
sulphur, the powdered assay [No. 77] is mixed
parts of a mixture of equal parts of potassiun
and sodium carbonate, the mass introduced i
sealed at one end, and heat applied, at first ve
but gradually raised to redness. A ring of metall
will be deposited in the colder part of the tube.

Fig. 31.

71. When *arsenic sulphides* are heated on cha
whole of the arsenic, especially when only sma
ties are present, may pass off in combination
phur; but when such compounds [No. 17] a
with from three to four parts of neutral pota

alate or potassium cyanide and exposed to the reducing flame, potassium sulphide is formed and the arsenic escapes with its peculiar odor, if not combined with cobalt or nickel.

72. To detect a very small quantity of *arsenic trioxide*, the following method may be used: a glass tube is closed by drawing it out to a point; the assay [No. 48] is introduced into the point and a charcoal splinter placed just above it; the tube is then heated to redness at the place where the charcoal splinter lies, and as soon as this is incandescent, heat is also applied to the assay. (Fig. 32.) If any moisture given off from the substance when

Fig. 32.

heated condenses upon the tube, it should be removed by means of a roll of filter-paper. The arsenic trioxide is volatilized, and its vapors, while passing over the red-hot charcoal, become reduced and deposit a black metallic ring of arsenic in the colder part of the tube. By cutting the tube below the ring and heating this part by the flame of a gas-lamp, the arsenic is volatilized, thereby emitting its characteristic odor.

73. To show the presence of arsenic in *arsenites* and *arsenates*, it will in most cases be sufficient to mix the substance [No. 48] with sodium carbonate and heat it on charcoal in the reducing flame. Sometimes it is necessary to treat the assay with a mixture of sodium carbonate and potassium cyanide in the manner mentioned (par. 70); and in other cases again, where but small quantities of arsenic trioxide or arsenic pentoxide are combined with metallic oxides which are readily reduced, recourse must be had to the humid way.

Bismuth. The reactions of bismuth and its compounds, see pars. 12, 17, 22, 26, 47, 155, Table II., 3.

74. Bismuth is usually detected by its coating. If in combination with sulphur and heated alone on charcoal, the yellow coating is surrounded by a white coating of bismuth sulphate. Heated with sodium carbonate, the sulphate is not formed. Bismuth, when alloyed with other metals, or when as sulphide in combination with other sulphides, is in many cases, and most especially so when accompanied by lead or antimony, not to be detected with certainty by the coating which it deposits on charcoal. In such a case the assay [No. 11] is treated on charcoal until a copious yellow coating is formed. The coating is carefully scraped off from the charcoal and dissolved in salt of phosphorus on platinum wire with the oxidizing flame. The colorless bead is removed from the wire, placed on charcoal, a little metallic tin added, and the whole exposed to the reducing flame. If bismuth was present, the glass assumes, on cooling, a dark-gray or black color. The antimony oxide showing the same behavior, the assay, if not quite free from antimony, has to be treated on charcoal in the oxidizing flame until the whole of it has been volatilized, and the remaining mass treated on another piece of charcoal as above mentioned.

According to Von Kobell, any compound of bismuth treated before the blowpipe with a mixture of equal parts of potassium iodide and sulphur on a large coal, gives a beautiful and very characteristic red coating, at quite a distance from the assay. In case of sulphide the sulphur is not necessary. If lead is present a deep-yellow coating is formed, which, however, does not interfere with the bismuth lying nearer the assay.

Cornwall * suggests the following method to detect bismuth in presence of lead and antimony: to the mixture of the three oxides an equal volume of sulphur is added, and the whole treated before the blowpipe in a deep cavity on coal with the blue flame for a short time. The resulting fused sulphides are removed to a flat coal and treated alternately with the oxidizing flame and reducing flame until antimony fumes have nearly ceased and an impure blue-lead flame appears. The residue is powdered and an equal part of a mixture of one part of potassium iodide and five of sulphur, by weight, added. This is then heated in an open tube, 10 to 12 cm. long and not less than 10 mm. wide, over a Bunsen gas-burner or spirit-lamp. A distinct red bismuth iodide sublimate is formed, about 10 mm. above the yellow sublimate of lead iodide.

Care must be taken not to confound with the bismuth sublimate, a sublimate of iodine, which may condense on the upper part of the tube, but at a greater distance from the assay.

75. Boric Acid. With many borates, which do not impart to the outer flame the ·peculiar yellowish-green color (par. 60), this reaction may be produced by reducing the substance [No. 33] to powder, adding a drop

* "American Chemist," March, 1872.

of concentrated sulphuric acid, fastening the mix
the hook of the platinum wire, and playing on it
blue cone of the flame (see par. 60).

76. Another way, and by which even a ve
quantity of boric acid in salts and minerals
detected, is: to reduce the substance to a very f
der, to mix it with from three to four parts of a
of four parts of acid potassium sulphate and
of fluorite, perfectly free from boric acid, and t
the whole with a little water into a thick past
mass is then fastened to a platinum wire and ex
the blue cone of the flame.   While the mass en
fusion boron fluoride is formed, which, on escap
ors the flame intensely yellowish-green.   The rea
pearing sometimes only for a few seconds, and th
tion being only momentary, the flame should
carefully observed.

The method proposed by Iles is extremely
The substance, finely powdered, is placed on
foil, moistened with sulphuric acid, the excess
volatilized by heat, and the powder then mad
paste with glycerine.   If this is taken up by a
wire and placed in the flame, the latter is colore
ish-green.

77. If borates of the alkalies or alkaline earth:
solved in dilute hydrochloric acid, and one end o
of turmeric paper dipped into the solution, after
100° C., this end of the paper will become *brow*
Hydrochloric acid alone, if too strong, will p
dark-brown on turmeric paper.   The reddish col
paper, if moistened with an alkali, is changed to

78. Bromine.   Bromides treated with salt of ph
and copper oxide on platinum wire, or treated

per sulphate on silver foil, show the same reaction as chlorides (par. 83), with this difference, that the blue color of the outer flame is rather greenish, especially on the edges [No. 39]. When the bromine is all driven off, the green flame of the copper alone remains.

79. To discriminate bromides from chlorides more distinctly, the bromide is fused with acid potassium sulphate, both in the anhydrous state, in a small matrass with long neck. Sulphur dioxide is evolved and the matrass is filled with yellow vapors of bromine, characterized by their peculiar odor. The color of the gas is only clearly seen by daylight. Silver bromide may be distinguished from silver chloride by the asparagus-green color which it assumes when exposed to the sunlight after fusion with acid potassium sulphate.

80. Goldschmidt, for the detection of a bromine compound alone, or in presence of iodine and chlorine, gives the following method: If a bromine compound is fused in an open glass tube with pulverized bismuth sulphide, made by fusing metallic bismuth with sulphur, a yellow sublimate is formed. An iodine compound treated in the same manner forms a red sublimate, and a chlorine compound a white one. With a little care these elements can be readily recognized in presence of one another.

The presence of iodine, on account of its violet vapors, often renders the bromine reaction somewhat uncertain.

Cadmium. The reactions of cadmium and its compounds, see pars. 12, 35, 47, 159, and Table II.. 4.

81. To detect a very small quantity of cadmium, 1 per cent. or less, in *sinc* or its *ores*, the pulverized assay is mixed with soda and exposed for a short time to the reducing flame on charcoal. A distinct coating of brown cadmium oxide is deposited. The zinc, being less vol-

atile, forms a coating only with continued blowing
16].

82. Chlorine. Some copper oxide is dissolve
means of the oxidizing flame in a bead of sodium
phate on platinum wire until the glass is nearly oj
some grains of the pulverized assay [No. 41] are
made to adhere to the bead, and both heated wi
tip of the blue cone of the flame.　If chlorine is p
the flame now assumes an intense azure-blue color,
to the formation of copper chloride (par. 61).
test is very delicate and will show the presence of
minute quantity of chlorine.　Bromine produces
llar but somewhat greenish (par. 78) flame (pars. 10

When a chloride is heated with dry potassium
mate and concentrated sulphuric acid, dark, browni
vapors of chloro-chromic acid are evolved, which
dense to drops of the same color.　With ammon
drate this liquid becomes yellow.

83. Another method is to place on silver-foil som
rous sulphate or some copper sulphate, to moisten i
a drop of water, and then to add the assay [No
After a while the silver will be found blackened.
stances which are insoluble in water have previor
be fused with a little soda on platinum wire, to
soluble chloride.

84. Chromium. Chromium oxide gives very
teristic reactions with the fluxes on platinum wir
Table II., 6), but when accompanied by a large qu
of iron, copper, or other substances which also int
color the borax and salt of phosphorus beads, the
mium color frequently becomes very indistinct.

85. In such a case, and when the chromium is
combination with silica, its presence may be detec

the following manner the assay-piece [No. 68] is reduced to a fine powder and mixed with about twice its
own volume of a mixture of equal parts of soda and
nitre. The mass is fastened into the hook of a thick
platinum wire, or placed into a small platinum spoon,
and treated with a powerful oxidizing flame. An alkaline chromate is formed which is dissolved in water, the
solution supersaturated with acetic acid, boiled, and a
crystal of lead acetate added. If chromium was present, a yellow precipitate of lead chromate will appear.
The precipitate may be collected on a filter and tested in
the borax and salt of phosphorus beads, when the characteristic chromium-reactions will be produced. If instead
of lead acetate, silver nitrate is added, a dark, purplish-
red precipitate of silver chromate is formed.

Silicates which contain only a little chromium, but
much iron, or other coloring oxides of metals, are not
decomposed by nitre. In this case a pulverized mineral
is fused on coal in the oxidizing flame with one part of
sodium carbonate and one-half to three-fourths parts of
borax to a clear bead; this is pulverized and evaporated
to dryness with hydrochloric acid. The chlorides thus
formed are dissolved in water, the silica filtered off, the
iron is oxidized by boiling with a few drops of nitric
acid, and the bases, sesquioxides of chromium, iron,
etc., precipitated by ammonia from the acid solution.
The precipitate is collected on a filter, washed, and
fused with soda and nitre as above. By this means alkaline chromates are formed, which can be decomposed
by acetic acid and lead acetate, as already described.

Cobalt. The reactions of cobalt, see Table II., 7.

86. To detect cobalt when in combination with other
metals, see pars. 47, 99, *c*, 164.

To show its presence in *arsenides*, the assay [No. 75] is placed on charcoal and heated until fumes of arsenic trioxide are no longer emitted. (Lead and bismuth, if present, form the characteristic coatings.) Borax is now added and the heat continued until the glass appears colored. If the color is not pure blue, the presence of iron is indicated. The glass is in this case removed from the globule, and the latter treated repeatedly with fresh quantities of borax until the pure cobalt-color is obtained. Nickel and copper, if present, do not enter into the flux before the whole of the cobalt is oxidized. If we wish to ascertain the presence of these metals, the glass which is colored by cobalt is removed from the globule, and the latter treated with fresh portions of borax in the oxidizing flame until the color of the bead becomes brown, indicative of nickel. The glass is again removed and the globule treated with salt of phosphorus in the oxidizing flame: when copper is present the bead assumes a green color, which remains unaltered on cooling. Treated with tin on charcoal, the glass turns opaque and red from the reduction of the suboxide.

To ascertain its presence in sulphides, the assay [No. 76] is heated on charcoal in the reducing flame until all sulphurous fumes are driven off; the remaining mass is then calcined to oxide, and the calcined mass is treated with borax in the oxidizing flame. When cobalt alone is present, the bead will be pure blue; but when iron is also present, the glass will not appear pure blue, but when copper and nickel are also present to some extent, their influence on the glass is slightly seen. The bead is in this case exposed to the reducing flame until it appears translucent and less green, the oxides of copper and nickel are by this means reduced, and the pure color of cobalt, or that of cobalt mixed with iron, becomes apparent. The separation of the metals may be promoted by adding a little pure lead, and the substance freed from an excess by treating it alone on coal, after which it is fused in the oxidizing flame with salt of phosphorus to detect nickel and copper.

Copper. The reactions of copper and its compounds, see pars. 47, 60, 61, 172, and Table II., 8.

88. The red color which copper imparts to the borax or salt of phosphorus bead, when heated on charcoal in the reducing flame in contact with tin (see Table II., 8), is very characteristic, and will in most cases clearly show the presence of this metal; but if only a small quantity of copper is associated *with other metals*, the reaction is not easily obtained; in this case we may proceed as follows:

The assay [No. 18, or No. 83, or No. 82] is placed on charcoal and played upon with the oxidizing flame until antimony and other volatile metals are driven off. Some vitrified boric acid is fused on charcoal to a glassy globule, the assay placed close to it, and the whole covered with a large reducing flame. When the metallic globule begins to assume a bright metallic surface, the flame is gradually converted into a sharply-pointed blue cone, which is made to act only on the glass, leaving the metallic globule untouched, and so situated that it touches the glass on one side, and on the other side is in close contact with the charcoal. During this process lead, iron, cobalt, part of the nickel, and such of the more volatile metals as were not entirely removed by the previous calcination, as bismuth, antimony, zinc, etc., become oxidized, and their oxides partly volatilized and partly absorbed by

8

the boric acid. The remaining metallic globule is then removed from the flux and treated on charcoal with salt of phosphorus in the oxidizing flame, when the copper is oxidized and dissolved. The limpid bead is then refused in the reducing flame with addition of tin. A trace of copper may thus be made to produce distinctly the characteristic reaction, rendering the cold bead distinctly red and wholly or partially opaque.

89. To show the presence of copper in compounds which contain much *nickel, cobalt, iron,* and *arsenic,* the assay [No. 79] is first treated with borax on charcoal in the reducing flame, when the greater part of iron and cobalt are dissolved. The remaining globule is then mixed with some pure lead and treated as shown in par. 88. Arsenic is for the most part driven off, and the rest of the iron and cobalt, with some nickel, absorbed by the boracid. The globule is removed from the glass and treated with salt of phosphorus in the oxidizing flame; dark-green while hot, and somewhat lighter green when cold (produced by the mixture of the yellow of nickel and the blue of copper), indicates the presence of copper.

To detect copper when in combination with tin, see par. 124.

90. To detect copper in *sulphides*, the pulverized assay [No. 73] is calcined and the calcined mass treated as above, or, when the amount of copper is not very small, simply treated with borax or salt of phosphorus on charcoal in the oxidizing flame, and subsequently with addition of tin in the reducing flame. The presence of copper is then shown by the red color and the opaqueness of the glass on cooling. This reaction is only prevented, or at least made indistinct, by antimony or bismuth, which cause the glass to turn gray or black. In this case the

assay is, after calcination, mixed with soda, borax, and some pure lead, and the mixture fused on charcoal in the reducing flame. The metallic globule is then heated on charcoal to drive off the antimony, and afterwards treated with boric acid as above.

91. When a mineral which contains copper is heated in the blue cone, the outer cone of the flame frequently assumes a green, or, if the metal is in combination with chlorine, an azure-blue, color. This reaction, if not produced by heating the substance alone, may sometimes be obtained by adding a drop of concentrated hydrochloric acid to the pulverized assay [No. 70], evaporating to dryness, mixing the dry powder with a little water to a stiff paste, fastening this into the hook of a platinum wire, and then exposing it to the blue cone of the flame.

92. Fluorine. To detect fluorine in those minerals where it occurs only as an accessory element in combination with weak bases, and which at the same time contain water, as in No. 57, a small piece is placed in a glass tube sealed at one end, a wet Brazil-wood paper introduced into the open end, and heat applied. Hydrofluoric acid is evolved, which turns the red color of the test-paper to straw-yellow and corrodes the glass. Mica, containing not more than ¾ per cent. of fluorine, shows the reaction very distinctly.

If the mixture is heated from above downward, it is not so likely to be thrown out from the tube. The etching of the tube is best seen after the tube is cleaned and dried.

93. Another process, and by which the presence of fluorine in all kinds of compounds may be shown, is to mix the pulverized assay with some salt of phosphorus which has previously been fused on charcoal and then reduced

to powder ; to place the mixture on platinum-foil, which
is connected with an open glass tube in such a manner as
to constitute a kind of tubular continuation to the former,
and to heat with the blowpipe flame until the mass enters
into fusion.   If the flame is so directed that the products
of decomposition are made to pass through the glass tube,
and a moistened Brazil-wood paper is introduced into the
other end, the presence of hydrofluoric acid is indicated
by the change of color which the latter experiences, and
often by its pungent odor.   In some cases the glass will
also be dulled or a deposit of silica be formed.   This
test is very delicate.

94. Gold.   (See pars. 30, 47, 170, and Table II., 10.)
When gold is in combination with metals which are
volatile at a high temperature—ex. gr. tellurium, mercury, antimony—it is only necessary to heat the alloy on
charcoal with the oxidizing flame, when the gold remains
behind in a pure state, and may be recognized by its physical properties.   Lead is removed by the process of cupellation, as explained in par. 117.

95.   When associated with copper, the presence of which
is easily detected by salt of phosphorus on charcoal, the
alloy—for example, gold coin—is dissolved in pure melted
lead and the new compound subjected to the process of
cupellation on bone-ash.   Copper is by this means entirely removed.   To test the remaining globule for silver,
it is treated with salt of phosphorus on charcoal in the
oxidizing flame ; the silver is gradually oxidized and dissolved by the glass, which when cold assumes an opal-like
appearance.   To determine approximately the relative proportions of the two metals, the metallic globule is taken
from the cupel, placed in a small porcelain dish containing some nitric acid, and heat applied.   If the alloy con-

tains 25 per cent. of gold or less, it turns black, the silver is gradually dissolved, and the gold remains behind as a brown or black spongy or pulverulent mass. If the alloy contains more than 25 per cent. of gold, the globule turns also black, but the silver is not dissolved. If both metals are present in about equal proportions, the globule remains unaltered. If the amount of gold is considerable, it is indicated by the color of the alloy.

In both of the latter cases it must be fused on coal with borax and at least twice its weight of silver, free from gold, and then treated with nitric acid, when the separation will be complete. To form a gold button, it must be well washed with distilled water and fused on coal with borax, and it will then have the pure gold color and bright surface.

96. When associated with metals which alone are infusible before the blowpipe—as ex. gr. platinum, iridium, palladium—the metallic globule obtained by cupellation shows much less fusibility than pure gold. The exact nature of the foreign metals cannot be ascertained before the blowpipe ; the humid way must be resorted to.

97. Iodine. Iodides, tested with a salt of phosphorus bead which is saturated with copper oxide, as shown in par. 82, impart to the outer flame a fine green color [No. 40].

Fused with acid potassium sulphate in a glass tube closed at one end, violet vapors are evolved, iodine sublimes, and sulphur dioxide is given off.

Iodides mixed with about one-third of their weight of copper sulphate and heated in a glass tube are decomposed, as shown by the violet vapors, which color paper, moistened with starch, blue.

98. Iodine, in combination with silver or with alkalies, can be detected in the presence of other halogens by mixing the powdered substance with bismuth sulphide (prepared by heating bismuth and sulphur together) and heating on charcoal before the blowpipe flame. A red coating of bismuth iodide is formed if iodine is present.

Iron. The reactions of the oxides of iron, see Table II., 13. Also pars. 12, 47, 162.

99. *a.* To distinguish protoxide from peroxide, the substance is added to a borax bead containing copper. With peroxide the bead is colored bluish-green, whilst with protoxide red lines or flakes of cuprous oxide appear.

b. To detect iron along with *easily-fusible metals*, such as lead, bismuth, antimony, tin, or zinc, the substance is heated on charcoal with borax in the reducing flame. The easily-reducible metals do not become oxidized, and consequently are not absorbed by the glass. The glass is separated from the metallic bead, and is heated on a fresh piece of charcoal in the reducing flame, when it acquires the characteristic bottle-green color produced by iron, and becomes vitriol-green on addition of tin.

c. In presence of *cobalt* the bead is not green, but blue in color. In such case iron is sought for by heating the blue glass on platinum wire in the oxidizing flame sufficiently long to convert all the iron into peroxide. With very little iron present, the bead is green when hot, and blue when cold ; with more iron, the bead is dark-green when hot, and pure green when cold, this latter resulting from a mixture of the yellow iron and blue cobalt colors. The residual metal on the charcoal after the treatment with borax (often only nickel and copper) is examined according to par. 88.

d. An admixture of *manganese* colors the bead in the oxidizing flame blood-red. By reduction with tin on charcoal the bead becomes vitriol-green. If *cobalt* be present along with *manganese*, a dark-violet bead is produced in the oxidizing flame, which in the reducing flame becomes green when hot and blue on cooling.

e. To test for iron in *nickeliferous* substances, the assay is dissolved in borax in the oxidizing flame and then heated on charcoal in the reducing flame. Metallic nickel separates out, and the iron, remaining dissolved in the glass, colors it green.

f. A substance containing *iron* and *copper* gives a green borax bead in the oxidizing flame both before and after cooling; from this bead copper separates on charcoal under the reducing flame, and the glass becomes green from iron. If the amount of copper present be small, the assay is fused together with borax, sodium carbonate, and assay-lead, the metallic bead obtained heated with boric acid in the oxidizing flame, and the copper sought for by the aid of salt of phosphorus and tin.

g. If *iron* and *chromium* occur together, the color of the glass affords no indication of the presence of iron. The substance is fused with sodium carbonate on charcoal in the reducing flame, the reduced iron is separated from the slag by washing, and the latter is fused with potassium nitrate for the detection of chromium.

h. *Iron* and *uranium oxides* cannot be distinguished from one another in the dry way. To separate them, the assay is fused with acid potassium sulphate, extracted with water, and the solution treated with ammonium carbonate to precipitate the iron; the filtrate is acidified, boiled to expel carbon dioxide, and the yellow uranium precipitated by ammonia. Both products are then further examined.

i. A substance containing *iron, nickel, cobalt, manganese,* and *copper* is fused with metallic arsenic or with potassium arsenate, and the mass is treated with borax in successive portions in the oxidizing flame.

There results—

First, a yellowish-green color from iron.
Then a blue   "   "   cobalt.
 "   a brown   "   "   nickel.
 "   a green   "   "   copper.

Under the reducing flame nickel and copper can be separated from the borax glass, whilst iron, cobalt, and manganese remain dissolved, and are looked for according to par. 99, *d*.

Lead. The reactions of lead and its compounds, see pars. 12, 27, 61, 158, and Table II., 15.

100. An alloy of lead and zinc [No. 12] deposits a coating of lead oxide mixed with zinc oxide; the presence of lead is shown by the color of the coating and by the azure-blue tinge which it imparts to the reducing flame (see par. 27). Test the zinc coating with cobalt solution, which turns it yellowish-green when heated.

An alloy of lead and bismuth [No. 11] deposits a coating somewhat darker than that of pure lead, in which the presence of bismuth may be detected as shown in par. 74, and the presence of lead by the azure-blue color of the reducing flame.

101. To detect lead in sulphides, the substance is placed on charcoal and treated with the reducing flame; the lead is detected by its coating. An admixture of antimony cannot by this means be ascertained, since the ring of lead sulphate surrounding that of the oxide bears a striking resemblance to the coating formed by antimony oxide. In this case the pulverized assay [No. 82] is mixed with a sufficient quantity of soda and treated for a short time with the reducing flame. If no antimony is present a pure yellow coating with bluish-white edges is formed; but in presence of antimony this coating is surrounded by another white one of antimony oxide. The lead oxide coating appears, moreover, darker than usual, resembling that of bismuth, owing probably to the formation of lead antimonate. A very small quantity of antimony by this method cannot be found out with certainty, since, by keeping up the blast for some time, the sodium sulphide begins to vaporize and to coat the charcoal with a ring of sodium sulphate (see par. 38).

102. When lead sulphide is associated with a considerable quantity of copper sulphide [No. 18], the metallic globule obtained by the process of reduction does not betray, by its physical properties, the presence of lead. But if the alloy is removed from the flux and played upon with a powerful oxidizing flame, the greater part of the lead will be volatilized and deposit a coating.

Lead chloride before the blowpipe first fuses and then gives two coats—one of the chloride, white and volatile, and another of the oxide, less volatile. It also imparts a blue color to the reducing flame.

Lead phosphate alone on coal fuses to a globule, and affords no coat or a very slight one. Crystallizes on cooling.

103. **Lithium.** To detect lithium in silicates which contain only little of it, proceed as follows: The substance [No. 64] is reduced to a fine powder and mixed with about two parts of a mixture of one part of fluorite with one and a half parts of acid potassium sulphate; a few drops of water are added and the whole kneaded into a paste. The mass is fused with the blue cone of the flame into the hook of a platinum wire. If lithia is present the outer flame will appear red. If only a small amount is

present the color is not very intense, and verges into violet. The presence of potassium does not prevent the reaction, but makes the flame appear still more violet; sodium makes the reaction uncertain.

If boric acid be present in the silicate, as in tourmaline, the outer flame at first exhibits a green tinge, but afterward a wine or less intense red from the lithia.

In presence of phosphoric acid —as in case of tryphylite, for example — it causes a green flame, perceptible along with the red one, especially after moistening with sulphuric acid.

Another method of detecting lithium when mixed with sodium is to dip the assay, moistened with hydrochloric acid, into melted wax, and then heat it in the blue flame, by which the red color is produced immediately.

Manganese. The reactions of manganese, see Table II., 16. Also pars. 47, 181.

104 If a bead containing manganese, just taken from the oxidizing flame, be brought into contact with a crystal of potassium nitrate or chlorate, or be thrown into a porcelain capsule containing the powdered reagent, a violet frothy mass of potassium permanganate is formed.

The presence of manganese in any compound substance is readily detected by mixing the pulverized assay [No. 63 or No. 81] with about two parts of soda and one of nitre, and fusing it by means of the oxidizing flame on platinum foil. Potassium manganate is formed, which, while hot, is green and transparent, and on cooling turns bluish-green and opaque. The slightest trace may be detected in this way. Chromium does not prevent the reaction, merely changing the color to yellowish-green. It is only in presence of silica and cobalt that this test is not available, since at a high temperature the silica unites with the soda to form sodium silicate, which, in dissolving the cobalt oxide, produces a blue glass, and thus interferes with the manganese color. In this case the silica must first be separated in the wet way.

Metallic compounds containing manganese should be dissolved in nitric acid, the solution evaporated to dryness, and the ignited residue tested with sodium carbonate and potassium nitrate, as above.

Mercury. The reactions of mercury and its compounds, see pars. 10, 11, 18, 47, 156, and Table II., 17.

105. Mercury is detected in amalgams [No. 9] by the sublimate of metallic mercury which they yield when heated in a glass tube closed at one end. The globules if small are best seen with a small magnifying-glass.

When in combination with sulphur [No. 78], chlorine [No. 49], iodine, or oxygen-acids, the substance is previously mixed with some anhydrous soda or some neutral potassium oxalate. The acids, etc. are retained by the soda, and mercury sublimes.

If the quantity of mercury is so small that the nature of the sublimate cannot with certainty be ascertained, the experiment has to be repeated, a piece of iron wire around which a gold-leaf has been wrapped being at the same time introduced into the tube and held close above the assay. The gold-leaf will turn white, even when the amount of mercury present is very small.

Molybdenum. For the reactions of molybdenum and its compounds, see pars. 38, 51, 60, and Table II., 18.

106. Small quantities of molybdic acid may be detected by adding a little of the powdered substance to some strong sulphuric acid on a piece of platinum bent up at the sides. After heating till evaporation begins and then cooling, the foil is repeatedly breathed upon. Where only blue spots occur on cooling, if breathed upon an intense blue color is produced. Or if a little alcohol be added to the blue spots instead of breathing upon them, and then burnt off, the color is produced (see par. 174).

Nickel. The reactions of nickel, see Table II., 19.

107. Fusible metallic compounds of nickel are treated with borax on charcoal in the reducing flame; iron, cobalt, etc. enter into the flux and may be detected as shown in par. 86, while the metals the oxides of which are easily reduced remain behind. This operation is repeated until the glass appears no longer colored. The remaining globule is treated with salt of phosphorus in the oxidizing flame. We now obtain either the pure color of nickel or that of nickel mixed with copper, yellowish-green (see par. 89); in this case it is treated on charcoal with tin, whereby the presence of copper may be ascertained, the bead becoming opaque and red. Bismuth or antimony prevents the reaction for copper, the bead turning black instead of red. Such compounds must, previous to their treatment with fluxes, be heated on charcoal in the reducing flame until all volatile substances are driven off [No. 79].

In arsenides and sulphides, nickel is detected by the methods given for cobalt under the same circumstances (see par. 87).

Small quantities of nickel in the presence of cobalt may be detected by treating a small quantity of the substance with borax on platinum wire; a dark-colored bead is formed, which is placed on charcoal with a small gold bead and fused in the reducing flame. When cold, the gold bead is separated from the slag by a slight blow with a hammer, and is fused with salt of phosphorus in the oxidizing flame. The glass takes up the easily-soluble

~~cobalt oxide, becoming~~ blue; and fresh quantities must ~~be added until the color~~ changes to green, and finally becomes yellow. The gold may afterward be refined by cupelling with lead on bone-ash (see par. 163).

108. Nitric acid. The perfectly dry substance [No. 43] is heated in a matrass with some acid potassium sulphate; orange-yellow vapors of nitrogen tetroxide are emitted, even if but a small quantity of a nitrate is present. Or if chlorine is present the substance should be heated with litharge free from lead peroxide, which at first absorbs the nitric acid, but yields it up at a higher temperature. A piece of paper moistened with a solution of ferrous sulphate, free from peroxide and acidulated with sulphuric acid, is inserted into the neck of the tube, which should be rather long, and nitrogen tetroxide if present will color the paper yellowish to brown (see par. 60).

109. Phosphoric acid. A very minute quantity of phosphoric acid may be detected by pulverizing the substance [No. 65], adding a drop of concentrated sulphuric acid, fastening the paste into the hook of a platinum wire, and playing upon it with the blue cone of the flame; the outer flame will assume a bluish-green color (see par. 60).

Certain nitrogen compounds, as nitric acid, ammonium nitrate, ammonium chloride, etc., when fastened into the hook of a platinum wire and touched with the cone of the blue flame, impart to the outer flame a bluish-green color resembling that caused by phosphoric acid (see par. 60).

110. For very small quantities of phosphoric acid Bunsen has proposed a test which consists in mixing the substance with two or three times as much soda, and placing the completely dried mixture in the drawn-out part of a small tube, similar to those used in testing for arsenic. The mixture is again heated to remove all moisture, a long bit of sodium, or, better, magnesium wire, inserted into it, and fused with the blowpipe. When cold, the portion of the tube containing the fused mass is broken off, placed in a porcelain dish, and wet with a few drops of water; if phosphoric acid was present, the phosphuretted hydrogen formed may be recognized by its odor.

111. **Potassium.** The violet color of the flame is sufficiently characteristic for potassium (see par. 41). But being altogether prevented, or at least made very indistinct, by the addition of a few per cent. of soda or lithia, it can only in a very few cases be made use of. For the detection of potassium in silicates it is almost entirely unavailable, because these compounds almost always contain some soda.

112. If the base of a compound consists essentially of potassium, the following method may be advantageously employed for its detection: Some borax, to which a little boric acid has been added, is melted into the hook of a platinum wire, and so much pure protoxide of nickel, free from cobalt, added that the glass on cooling shows a distinct brownish color. A small piece of the substance under examination [No. 38] is made to adhere to the glass, and the whole fused together with the oxidizing flame. If the assay-piece contained no potassium, the color of the glass, after perfect cooling, will have remained unchanged; but if potassium was present in sufficient quantity, the glass will appear bluish.

The simplest means of detecting potassium in a salt in which, owing to a greater or less amount of soda, the violet coloration of the flame cannot be recognized, consists in viewing the color of the flame through deep-blue cobalt glass or a stratum of indigo solution (see page 119). The presence of potassium is recognized, according to the thickness of the intervening medium, by the violet or poppy-red color, while a very large amount of soda produces a blue color, and a smaller quantity is not perceptible.

The carbon of organic matter produces the same color as potassium, and if contained in the assay should be removed by ignition.

If lithium is present, a thicker stratum of solution or darker glass must be used.

According to Merz, a green glass may be used in some cases with advantage, the lithium flame being invisible through it, while the potassium and barium flames appear bluish-green, and that of sodium orange-yellow.

In testing silicates with the cobalt glass, they should first be heated with pure gypsum in the flame, thus forming sulphate of the alkali, which is volatile, and imparts to the flame its characteristic color.

113. Selenium. The reactions of selenium are very characteristic. In non-volatile compounds, which do not give the red sublimate mentioned in par. 11, γ, the selenium is detected by heating a small piece of the substance [No. 84] on charcoal in the oxidizing flame, when the peculiar odor is evolved ; if much selenium is present, a coating is deposited (see par. 36). Selenites and selenates are treated on charcoal with soda in the reducing flame, when a reduction takes place and the selenium vaporizes with the characteristic odor (see pars. 20, 61, 152).

114. Silica. Pure silica [No. 51], when treated with borax on platinum wire, dissolves slowly to a transparent glass which fuses with difficulty. Treated with salt of phosphorus in the same manner, only a small quantity is dissolved, the rest floating in the liquid bead as a semi-transparent mass. The behavior with soda, see par. 45. With a little cobalt solution it assumes a pale-bluish

, which, on addition of a large quantity of the re-
, turns dark-gray or black; very thin splinters may
sed by a great heat to a reddish-blue glass.

, Silicates [No. 58], when treated with salt of phos-
is on platinum wire, are decomposed; the bases unite
the free phosphoric acid to a transparent glass, in
ı the silica may be seen floating as a gelatinous,
ly mass. The bead ought to be carefully observed
: hot, since many silicates form a glass which on
ng opalizes or becomes opaque, when, of course, the
omenon can no longer be seen. The experiment is
performed with a small splinter of the substance un-
kamination, and only when this does not appear to be
ed by the flux, the finely-pulverized substance should
ed. If but a very small quantity of silica is present,
lass will appear perfectly transparent. Its presence in
ase cannot be detected by means of the blowpipe.

, Silicates containing at least so much silica that the
ity of oxygen in the acid is twice that of the oxygen
ı base, dissolve, when treated with soda on charcoal,
effervescence, to a transparent glass, which remains
cold. When less silica is present decomposition
takes place, but the glass turns opaque on cooling,
mount of sodium silicate which is formed not being
ient to dissolve the eliminated bases.

ver. The reactions of silver, pars. 29, 171, and Table
7.

, When in combination with metals which are vol-
at a high temperature—for example, bismuth, lead,
antimony—the substance is heated alone on charcoal,
, after volatilization of these metals by long blow-
ı button of pure silver remains behind, and a red-
coating is deposited on the charcoal. If associated

with much lead or bismuth, these metals are best removed by cupellation, a process which is performed in the following manner: Finely-pulverized bone-ash is mixed with a minute quantity of soda, and made with a little water into a stiff paste; a hole is now bored into the charcoal, filled with the paste, and its surface smoothed and made slightly concave by pressing on it with the pestle of the little agate mortar. The mass is then dried by the flame of a gas or spirit lamp. On this little cupel the assay [No. 13] is placed, and heated with the oxidizing flame until the whole of the lead or bismuth is oxidized and absorbed by the cupel. The silver, or if gold is present the alloy of silver and gold, remains as a bright metallic button on the cupel.

118. When combined with metals which are not volatile, but which are more easily oxidized than silver, the presence of this metal may in some cases be detected by simply treating the alloy with borax or salt of phosphorus on charcoal. Copper, nickel, cobalt, etc. are oxidized, and their oxides dissolved by the flux, while silver remains behind with a bright metallic surface. But when these metals are present to a considerable extent, another course has to be pursued—a course which may always be taken when a substance is to be assayed for silver or silver and gold.

119. The assay-piece [No. 83] is reduced to a fine powder, mixed with fused borax and metallic lead (the quantities of which altogether depend upon the nature of the substance, and for which, therefore, no general rule can be given), and the mass placed in a cylindrical hole of the charcoal. A powerful reducing flame is given until the metals have united to a button, and the slag appears free from metallic globules. The flame is now con-

9 *

verted into an oxidizing flame and directed principally upon the button.   Sulphur, arsenic, antimony, and other very volatile substances are volatilized; iron, tin, cobalt, and a little copper and nickel become oxidized and are absorbed by the flux; silver and gold and the greater part of the copper and nickel remain with the lead (and bismuth, if present).   When all volatile substances are driven off, the lead begins to become oxidized and the button assumes a rotary motion; at this period the blast is discontinued, the assay is allowed to cool, and when perfectly cold the lead button is separated from the glass by some slight strokes with a hammer.   It is now placed on a cupel of bone-ash and treated with the oxidizing flame until it again assumes a rotatory motion.   If much copper or nickel is present, the globule becomes covered with a thick infusible crust, which prevents the oxidation; in this case another small piece of pure lead has to be added.   The blast is kept up until the whole of the lead and other foreign metals—viz. copper and nickel—are oxidized; this is indicated by the cessation of the rotatory movement, if only little silver is present, or by the appearance of all the tints of the rainbow over the whole surface of the button if the ore was very rich in silver; after a few moments it takes the look of pure silver.   The oxides of lead, copper, etc. are absorbed by the bone-ash, and pure silver, or an alloy of silver with other noble metals, remains behind; the button may be tested for gold, etc. after the method given in par. 95.

The silver chloride can be reduced on coal with soda.

120. **Sulphur.**   The presence of sulphur in sulphides may in many cases be detected by heating in a glass tube (see pars. 10, 11), or on charcoal with the oxidizing

In testing for sulphur, an alcohol or other flame free from sulphur compounds, and not coal gas, should be used (see pars. 49, 185).

121. A very delicate test for the presence of sulphur, in whatever combination it may be contained in the substance, and which possesses, moreover, the advantage over all other methods of being very easily performed, is to mix the pulverized assay [No. 36] with some pure soda, or, better still, with a mixture of two parts of soda, perfectly free from sulphates, and one of borax, and to treat it on charcoal with the reducing flame. The fused mass is removed from the charcoal, powdered, the powder placed on a silver foil or a bright silver coin, and a drop of water added. If the substance under examination contained any sulphur, a black spot will be formed on the silver foil, owing to the formation of silver sulphide from the decomposition of the sodium sulphide, which in its turn resulted from the decomposition of the sulphide or sulphate, or other sulphur compound of the assay-piece, under the influence of soda, charcoal, and a high temperature. Selenium and tellurium show the same reaction. The former is readily recognized by the peculiar odor which it emits when heated on charcoal alone, and the latter by its coating and flame.

If a substance containing sulphur is fused with sodium carbonate in the reducing flame, moistened with water in a watch-glass, and a little sodium nitro-prusside added, a fine reddish-purple color is produced.

A dilute solution of ammonium molybdate with an excess of hydrochloric acid is colored fine blue by a small quantity of sulphuretted hydrogen or sulphides dissolved in water.

Sulphides treated with hydrochloric acid liberate sulphuretted hydrogen, which may be recognized by its color and blackening a piece of paper moistened with lead acetate.

To decide whether the reactions obtained in the experiments above were owing to the presence of a *sulphite* or to that of a *sulphate*, the finely-pulverized substance [No. 76] is fused in a small platinum spoon with some potassium hydrate. The spoon with the contents is then placed in a vessel containing some water, and a piece of silver foil placed in the liquid. If the silver remains perfectly bright, a sulphate was present; if it turns black, a sulphide. The absence of substances which might exercise a reducing influence is required.

122. **Tellurium.** The presence of tellurium in mineral substances is detected by the tests given in par. 11, 37, 61. In presence of *lead* or *bismuth* the reactions in the open tubes and on charcoal are not quite sure. In this case we may subject the assay to the following treatment: The substance is mixed with some soda and charcoal-powder, the mixture introduced into a glass tube closed at one end and heated to fusion; after cooling, a few drops of hot water are poured into the tube; if tellurium was present, sodium telluride has been formed, which dissolves in hot water with a purplish-red color. This test is applicable to show the presence of tellurium in a great many compounds, even when it occurs in the oxidized state.

Natural tellurium compounds, when gently heated in a mattrass with an excess of sulphuric acid, impart to it a purple or hyacinth-red color, which disappears on adding water, while a blackish-gray precipitate is formed. When a mineral containing tellurium is treated on coal it generally yields a white tellurium dioxide coat, with a reddish-yellow border, which disappears under the reducing flame, imparting to the flame a green, or, in presence of selenium, a bluish-green, tinge. The horse-radish odor would be a certain indication of selenium.

If the mineral contains *lead* or *bismuth*, and is treated alone on coal for only a few moments, no pure tellurium dioxide coat is obtained, but a mixture of this with lead or bismuth oxide is liable to be deposited. This difficulty can be remedied by mixing the powdered assay with an equal volume of vitrified boric acid and treating it in the reducing flame. The lead or bismuth oxide is dissolved in the boric acid, notwithstanding the reducing flame, and yields no coat, while the tellurium alone volatilizes and coats the coal. When much selenium is also present a portion of it is deposited on the coal, and then the tellurium dioxide coat is less distinct. In such cases the mineral must also be tested in the open tube.

Tin. The reactions of tin and its compounds, see pars. 12, 28, 55, 173, and Table II., 30.

123. The presence of tin is indicated by its coating when the substance [No. 7], alone or mixed with soda, is exposed to the reducing flame on charcoal.

If a small quantity of a tin compound be added to a borax bead colored blue by copper oxide, and the reducing flame be applied, the bead becomes brown.

If substances containing tin oxide are heated on charcoal with soda and borax in the reducing flame, malleable beads of tin are obtained. These are separated from the slag and heated in the oxidizing flame, which converts them into white oxide, and is deposited on the charcoal near the assay. Treated with cobalt solution, the coating becomes bluish-green. When occurring together with zinc it can only be detected with certainty in the wet way.

124. To detect *copper in tin or its alloy—as bronze,* bell-metal, and gun-metal—the assay [No. 15] is fused with a flux consisting of one part of soda, one-half part of fused borax, and one-third part of silica. The flame is so directed that the metallic globule assumes a rotatory motion. When in this state the glass is kept covered as much as possible with the oxidizing flame, care being taken that the globule is at one side in contact with the glass, and at the other with the charcoal. The tin becomes oxidized, and the oxide, in a measure as it is formed, absorbed by the flux, whilst the copper remains behind. The latter is separated from the glass and further treated with salt of phosphorus, whilst the slag is powdered and reduced on charcoal with sodium or potassium carbonate.

Tin, when present in alloys, is almost always detected on fusing them upon coal; the globule is crusted with oxide, which can be removed with some difficulty after adding borax.

Sulphides containing tin, but forming no coat of oxide of tin near the assay, when treated alone on coal, must be roasted and treated in the reducing flame with soda and borax, when metallic tin is obtained, which may be tested alone on coal. If other reducible metals are present they form an alloy, in which the other metals can be recognized by means of the fluxes.

Titanium. The reactions of titanium are given in pars. 45, 51, 55, 176, and Table II., 31.

126. Titanium dioxide, when forming the principal constituent of any mineral substance, is easily detected by its behavior with the fluxes, but when in combination with bases these reactions are not always clearly perceptible,

ng frequently obscured by the predominating reaction

of the base. In such cases we may subject the assay to the following treatment, by which even very small quantities of titanium dioxide will become apparent: The substance [No. 62] is reduced to a very fine powder, mixed with from six to eight parts of acid potassium sulphate, and fused in a platinum spoon at a low red heat; the fused mass is dissolved in a porcelain vessel in the smallest possible quantity of water, aided by heat. If concentrated, it may be heated to boiling. There remains an insoluble residue, which is allowed to settle; the clear liquid is poured off into a larger vessel, mixed with a few drops of nitric acid and at least six volumes of water, and heated to ebullition. If the substance under examination contained any titanium, a white precipitate of metatitanic hydrate forms on boiling. If the solution is not acidified with nitric acid before boiling, a yellow, ferruginous precipitate is obtained when the substance contains iron. The precipitate is collected on a filter, washed with water, acidulated with nitric acid, and tested with salt of phosphorus, either on platinum wire or on coal. If the amount of metatitanic hydrate is so small that it does not give in the reducing flame to the salt of phosphorus the violet color of titanium dioxide, it is only necessary to add a little iron sesquioxide when the assay is upon a wire, or a small piece of iron wire when on coal, and to fuse the glass for a short time with the reducing flame; it appears yellowish while hot, and brownish-red when cool.

If titanium dioxide be fused with caustic potash, dissolved in water, and the solution evaporated after addition of an excess of hydrochloric acid and a piece of tin-foil, the liquid becomes violet-colored, and, on dilution with water, rose-red.

Tungsten. The reactions are given in pars. 45, 51, 175, and Table II., 32.

126. Tungsten may be detected by fusing the assay with five times its weight of sodium carbonate, the mass extracted with water, and the tungstic acid precipitated with hydrochloric acid in the form of a white powder. The precipitate becomes yellow on boiling, and is insoluble in excess of the acid (distinction from molybdic acid), but dissolves in ammonium hydrate. The solution, after acidification, gives a deep-brown coloration with potassium ferrocyanide, and after some time a precipitate of the same color separates; with silver nitrate a white, and with stannous chloride a yellow, precipitate is produced. On acidifying with hydrochloric acid and warming, the precipitate changes to a clear blue color, which is very characteristic.

127. **Uranium.** The presence of this metal is easily recognized, in substances which contain no other coloring constituents, by the reactions given in Table II., 33; the most characteristic test is that with salt of phosphorus. In presence of much *iron* this reaction becomes indistinct; we may then operate in the following manner: The finely-pulverized substance [No. 67] is fused with acid potassium sulphate, the fused mass dissolved in water, mixed with ammonium carbonate in excess, the liquid separated from the precipitate by filtration, and the filtrate heated to ebullition. If any uranium was present, a yellow precipitate is thrown down, which gives with the fluxes the reactions of pure uranium.

If the substance contains *copper* oxide, it is treated with soda, borax, and a silver bead on coal in the reducing flame until all the copper is reduced and taken up by the silver, after which the slag, containing uranium and other non-reducible oxides, like iron in a low state of oxidation, is dissolved by warming it with a little nitric acid, treated with excess of ammonium carbonate, and the process conducted as above (see par. 182).

Vanadium. For the reactions of vanadium and its compounds, see pars. 41, 51, 180, and Table II., 34.

128. On fusing vanadium compounds with soda and nitre on a platinum spiral, extracting with water, adding acetic acid in excess, and then silver nitrate, a yellow precipitate is formed. By evaporating the fused mass with aqua regia, a yellow or brownish solution is formed, which turns blue on the addition of stannous chloride.

If the solution of the fused mass in water is acidified and well shaken with hydrogen peroxide, it becomes red, and retains this color on the addition of ether, the latter remaining uncolored.

Zinc. The reactions for zinc and its compounds, see pars. 12, 34, 54, and Table II., 35. Also par. 160.

129. A small amount of zinc, when associated with considerable quantities of lead, or bismuth, or antimony, or tin, cannot always with certainty be ascertained by means of the blowpipe.

If the substance under examination contains the zinc as oxide [No. 22], or but a small quantity of sulphide, it is mixed with soda and treated on charcoal in the reducing flame. Substances consisting essentially of zinc sulphide may be thus treated without the addition of soda, and such as contain, besides zinc oxide, other metallic oxides, are conveniently mixed with some soda to which about one-half of its weight of borax has been added. A ring of zinc oxide is deposited on the charcoal. When lead is present [No. 12] the coating is frequently not pure, being mixed with the coating of lead.

10

In this case it is moistened with some cobalt solution and heated again with the oxidizing flame ; the lead oxide is reduced by the red-hot charcoal and volatilized, while the zinc oxide remains behind with a green color (see par. 56). It is well to moisten the charcoal at the point where the coating is to be formed with the cobalt solution.

The above reaction is not affected by the presence of lead and bismuth, but in the presence of much antimony a little zinc can only be found with difficulty before the blowpipe, for the oxides of antimony formed will have a green color, and cannot be driven off with the oxidizing flame. In many compounds, however, all of the antimony may be volatilized with the oxidizing flame, and the zinc then treated with the cobalt solution.

If tin is present the zinc cannot be recognized by the coating on coal, as its oxide assumes a bluish-green color with the cobalt solution.

When zinc occurs in small quantity it may be determined by using as a flux a mixture of nitre, sodium chloride, and soda, a coating being formed more readily from the volatile chlorides.

130. CONDENSED VIEW OF BLOWPIPE REACTIONS.

Arranged according to the phenomena observed.

EXPERIMENT.	OBSERVATION AND INFERENCE.
(1.) In Glass Matrass, or tube closed at one end.	*Water is given off* : water of crystallization; alkaline reaction = NH_3; acid reaction = volatile acids.
	Gas is evolved : O, SO_2, H_2S, NO_2, CO, CO_2, CN, NH_3, HF, I, Br.
	Sublimate is formed : white = NH_4Cl, Hg_2Cl_2, HgCl, Sb_2O_3, As_2O_3, TeO_2; grayish-black = As, Hg, Te; colored: *hot*, yellowish-brown; *cold*, yellow = S; *hot*, black; *cold*, reddish-brown = antimony sulphide; *hot*, brownish-red; *cold*, reddish-yellow = arsenic sulphide; black, on rubbing red = cinnabar; reddish-black (dark-red powder) = Se.
	Change of color : white, yellow, white = ZnO; white, yellow, brown, light-yellow = SnO_2; white, brownish-red, yellow = PbO; white, orange, pale-yellow, Bi_2O_3; red, black, red (volatile) = HgO; red, black, red (non-volatile) = Fe_2O_3.
	Substance melts : alkaline salts.
	Carbonization takes place : organic bodies.
	Substance becomes phosphorescent : alkaline earths, earths, ZnO, SnO_2.
	On heating with sodium carbonate, ammonia is evolved : ammoniacal salts and organic nitrogen compounds.

BLOWPIPE REACTIONS.—Continued.

EXPERIMENT.	OBSERVATION AND INFERENCE.
(2.) In the Open Tube.	*Odor emitted*: of SO_2 = sulphur and sulphides; of rotten horse-radish = selenium and selenides (steel-gray sublimate with red border). *Sublimate is formed*: white and crystalline = As; metallic globules = Hg; white fumes, sublimate partially volatile = Sb; white fumes, sublimate fusible to colorless drops = Te.
(3.) Alone on Charcoal, or Aluminium-foil.	*Substance is fusible*: (*a*) non-metallic bodies = alkaline and some alkaline-earthy salts; (*b*) metals: readily fusible = Sb, Pb, Cd, Te, Bi, Zn, Sn; difficultly fusible = Cu, Au, Ag. *Substance is infusible*: (*a*) non-metallic bodies = salts of the earths and alkaline earths, SiO_2; (*b*) metals = Fe, Co, Ni, Mo, Pt, Ir, Rh, Os, W. *Substance detonates*: nitrates, chlorates, bromates, and iodates. *Substance intumesces*: substances containing water, borates, alum. *Odor emitted*: S, As, Se, Te. *Flame-coloration*: yellow = Na; yellowish-red = Ca; red = Li, Sr; green = Ba, B_2O_3, P_2O_5, MoO_3, Cu; blue = CuCl, Se, As, Pb; violet = K.

BLOWPIPE REACTIONS.—*Continued.*

EXPERIMENT.	OBSERVATION AND INFERENCE.

(3.) Alone on Charcoal, or Aluminium-foil. (*Continued.*)	*Metal-reduction and coating :*
	
	(1.) *Metal reduced without coating.*
	
	  Shining bead = Ag, Au, Cu ;
	  Gray powder = Fe, Co, Ni, Mo, W, Pt, Pd.
	
	(2.) *Metal reduced with coating.*
	
	  coating, bluish-white and volatile = Sb ;
	  " *hot,* orange ; *cold,* lemon-yellow = Bi ;
	  " *hot,* lemon-yellow ; *cold,* sulphur-yellow = Pb ;
	  " *hot,* yellowish ; *cold,* white = Sn ;
	  " dark-red = Ag.
	
	(3.) *Coating without reduced metal.*
	
	  oating, white and volatile = As ;
	  " *hot,* yellow ; *cold,* white = Zn ;
	  " ‑ n, with variegated border = Cd ;
	  " steel-gray = Se ;
	  " white with red border = Te ;
	  " *hot,* yell w ; *cold,* white ; by blowing upon, first blue, then dark-red = Mo. (In oxidizing flame only.)

BLOWPIPE REACTIONS.—Continued.

EXPERIMENT.	OBSERVATION AND INFERENCE.
(6.) On Charcoal with Sodium Carbonate.	*Sodium sulphide formed:* sulphates and sulphides.
(7.) On Platinum-foil with Sodium Carbonate and Nitre.	*Metal-reduction and coating,* as in No. 3. Yellow mass = Cr. Green " = Mn.
(8.) On Charcoal with Cobalt Solution.	Blue infusible mass = Al_2O_3, SiO_2, and earthy phosphates and silicates. Blue glass = alkaline phosphates, borates, and silicates. Green mass = ZnO, TiO_2, SnO_2, Sb_2O_3. Flesh-colored mass = MgO. Brown mass = BaO. Gray mass = BeO, CaO, SrO. Not mass = ZrO_2.
(9.) With $Na_2S_2O_3$ in a tube closed at one end.	White = ZnO. Red = Sb_2O_3. Yellow = As_2O_3, CdO. Brown = SnO_2. Green = Cr_2O_3, MnO. Black = PbO, Fe_2O_3, CoO, CuO, NiO, UO_3, Bi_2O_3, Ag_2O, HgO.
(10.) With H_2SO_4, or $HKSO_4$ in Glass Matrass.	*Colored vapors:* brown = N_2O_4; yellowish-green = Cl_2O_4; violet = I; reddish-brown = Br. *Odorous gases:* SO_2, HCl, HF, H_2S. *Colorless and inodorous gases:* CO_2, CO.

BLOWPIPE REACTIONS.—*Continued.*

EXPERIMENT.	OBSERVATION AND INFERENCE.				
	Color of Bead.	Oxidising Flame.		Reducing Flame.	
		Hot.	Cold.	Hot.	Cold.
(5.) On Platinum Wire with Salt of Phosphorus.	Yellow (non-saturated bead).	Fe, U, V, Ce, Ag.........	V, Ni...........	Ti, Fe........	Ni.
	Yellow (saturated bead)......	U, V, W, Ti, Ag, Pb, Bi, Sb, Zn, Cd..	V, Ni...........	Ti............	Ni.
	Red (non-saturated bead) ...	Ni, Cr............	....	Cr..............	Cu.
	Red (saturated bead)........	Fe, Ce, Ni, Cr..	Fe (brownish).	Fe, Cr, Ni...	Cu.
	Brown.................	....	....	V.	
	Violet.................	Mn...............	Mn.............	....	Ti.
	Blue..................	Co..............	Co, Cu........	Co............	Co, W.
	Green.................	Cu, Mo........	U, Cr, Mo.....	U, Cu, Mo, W..........	U, Mo, Cr, V.

Silica Skeleton = SiO_2.

CHAPTER IV.

COLORED FLAMES, FLAME REACTIONS, AND
SPECTRUM ANALYSIS.

MANY substances, when brought into a colorless or non-luminous flame, color it in a remarkable manner. These colorations are, in many cases, characteristic of the elements yielding them, and furnish excellent means of detecting the latter, even in the minutest quantities, with great ease and certainty. Thus sodium-salts tinge the flame yellow; potassium compounds, violet; lithium salts, carmine-red; and on account of this peculiarity they may be distinguished from each other by the simplest experiments.

The Bunsen lamp, with chimney, previously described (Fig. 4), is especially adapted to such observations. The substance to be tested is brought by means of the platinum wire-loop (Fig. 12) into the zone of fusion of the gas-flame. The alkalies and alkaline earths are most remarkable in their coloring effects on the flame. If we compare together various salts of the same base, we find that they all, if volatile at the temperature of the flame, give the same color, but the color differs in intensity, being strongest with the most volatile salts, and *vice versâ*. Thus, potassium chloride gives a deeper tinge to the flame than potassium carbonate, and carbonate a stronger than potassium silicate. Sometimes a non-volatile compound is made to exhibit a characteristic tint by

118

Blowpipe Reactions.—*Continued.*

Experiment.	Observation and Inference.
(11.) With Zn and HCl after previous decomposition.	Solution is colored blue, green, blackish-brown = MoO_3. blue, copper-red = WO_3. blue, green, violet = V_2O_5. green = CrO_3. violet = TiO_2.
(12.) Heated on Platinum Wire, after Moistening with HCl.	*The flame is colored:* Yellow = Na; through blue glass invisible to blue; through green glass, orange. Yellowish-red = Ca; reaction obscured by Ba; through blue glass, greenish-gray; green glass, siskin-green. Red = Li; obscured by Na; through blue glass, violet; green glass, invisible. Red = Sr; obscured by Ba; through blue glass, violet-red; green glass, momentarily yellowish. Green = B_2O_3, P_2O_5, Ba (through green glass, bluish-green), MoO_3, Tl. Blue = Cu (subsequently green), In, Se, As, Pb, Sb. Violet = K; obscured by Na; through blue glass, purple-red; green glass, bluish-green.

these substances experiences ~~what the~~
after passing through ~~gradually thicker~~
solution.

131. The apparatus for these observations

1. A hollow prism, made of plate glass (
whose principal section forms a triangle, 1
of 150 millimètres and one of 35 millimètr
solution with which it is filled is prepared
1 part of indigo in 8 parts of fuming oil of
1500 to 2000 parts of water, and filtering.

In the following experiments the prism
izontally before the eye, so that the rays of
ways pass through gradually thicker layers o
The alkaline substances, brought singly inte
space, exhibit the following changes:

a. Chemically pure calcium chloride
duces a yellow flame, which, even with ve
of the indigo solution, passes through a t
Into the original blue-lamp flame.

b. Chemically pure sodium chloride, N

c. Chemically pure potassium carbona
potassium chloride, KCl, appears of a
violet, and at last of an intense crimson-re
seen through the thickest layers of solution
of sodium or calcium do not hinder the rea

d. Chemically pure lithium carbonat
lithium chloride, LiCl, gives a carmi
which, with increasing thickness of the
comes gradually feebler, and disappears be
est layers pass before the eye. Calcium a1
also without influence on this reaction.

2. A blue, a violet, a red, and a green gl
Is colored by cobalt protoxide; the violet,

sesquioxide; the red (partly colored and partly uncolored), by cuprous oxide; and the green, by iron sesquioxide and cupric oxide. The stained glasses found in commerce and employed for ornamenting windows generally possess the requisite shades of color.

Merz, who has made a complete investigation of this subject, employs with these glasses Bunsen's burner, and also a flame of pure hydrogen. The substances which he describes as giving characteristic colors to the flame of Bunsen's burner, in addition to those previously known, are nitric and chromic acids, while phosphoric and sulphuric acids give a peculiar coloration to the dark core of the flame of hydrogen.

132. The flame of Bunsen's burner gives three sorts of color:

a. **Border colors.** These are of course peculiar only to the most volatile substances. To produce them, the loop of platinum wire is to be held outside of the flame about one or two millimètres from the lower portion of the outer limit.

b. **Mantle colors.** Those, namely, which are seen when the substance is held in the bright blue-colored mantle which forms the outer portion of the flame.

c. **Flame colors.** To produce these, the loop is to be held horizontally and in the hottest part of the mantle. The hydrogen flame yields another species of color—viz the

d. **Core colors.** These are produced only by sulphuric and phosphoric acids, which communicate respectively a blue and green tinge to the cold core of the hydrogen flame.

The following, according to Merz, is a list of the more commonly-occurring substances which color the flame, with the color they impart:

11

BLUE FLAMES.
(*Consult page* 123.)

Intense blue, afterward green . . . . . . Copper chloride.
Pale clear blue . . . . Lead.
Light blue . . . . Arsenic.
Greenish-blue . . . . Antimony.
Blue mixed with green . . . . Copper bromide.
Blue core color . . . . Sulphuric acid.
Indigo-blue . . . . Indium.

GREEN FLAMES.
(*Consult page* 123.)

Bronze-green border color . . . . Nitric and nitrous acids.
" " " " . . . . . . Ammonium compounds.
" " " " . . . . . . Cyanogen "
Greenish-blue border color . . . . Hydrochloric acid.
Green mantle color . . . . Boric acid.
Gray yellow-green border color . . . . . Phosphoric acid.
Yellowish-green flame color . . . . . . Barium compounds.
Dark-green . . . . Iron wire.
Full green . . . . '. Copper.
Intense emerald-green . . . . " iodide.
Emerald-green, mixed with blue . . . . " bromide.
Pale-green . . . . Phosphoric acid.
Intense whitish-green . . . . Zinc.

RED FLAMES.
(*Consult page* 124.)

Intense-crimson . . . . Strontium compounds.
Reddish-purple . . . . Calcium "
Violet . . . . Potassium "
Dark brownish-red border color ⎞
and a rose-red mantle color   ⎠ . . . . Chromic acid.

YELLOW FLAMES.
(*Consult page* 125.)

Yellow . . . . Sodium compounds.
Feeble brownish-yellow . . . . Water.

133. Blue flames. Cuprio chloride gives an azure-blue zone, and **cupric nitrate** a pure green flame color. By the combined observation of both colors, **copper** may be distinguished from all other metals which give similar colors. The other flame-coloring metals—such as **arsenic, antimony, tin, lead, mercury,** and **zinc** —exhibit, especially in the form of chlorides, more or less intense bluish or greenish mantle colors, which, however, cannot be advantageously used as reactions for the metals themselves.

Sulphuric acid produces a beautiful blue core color, being reduced to **sulphur dioxide.** The free acid gives the color when the platinum loop is held in the border of the flame, but a sulphate must be held in the middle of the flame. In the latter case it is well to dip the test into strong **hydrochloric acid** or hydrofluosilicic acid.

134. Green flames. Nitric and nitrous acids give a bronze-green border color, usually with an orange-colored border. The test is to be previously dried in the flame, and dipped into a solution of **acid potassium sulphate,** or into dilute **hydrochloric acid,** according as we wish to test for nitric or nitrous acid. Ammonium and cyanogen compounds give the same bronze-green border, but more faintly. **Hydrochloric acid** gives a very weak greenish-blue border color, which lasts for a very short time, and therefore does not deserve attention. The acid is, however, decomposed, and the **chlorine** may easily be recognized.

Boric acid gives a beautiful green mantle color, which is so intense that the acid may be recognized in the presence of large quantities of phosphoric acid. Borates are to be decomposed with **sulphuric acid.** Phosphoric acid gives a gray yellow-green border color, as well as a beautiful green core color. The dry test is to be dipped into **sulphuric acid,** and held in the flame in the manner already pointed out, in order to show the border color. The green core color is less sensitive, but indispensable in recognizing phosphoric acid in the presence of large quantities of boric acid, and is produced by alternately moistening the test with a solution of hydrofluosilicic acid, and igniting it in the hydrogen flame, until the color distinctly appears.

Barium may be recognized by the yellowish-green flame color, which appears blue-green through the green glass. If the green disappears, and a red flame color makes its appearance, the test is to be repeatedly moistened with **hydrochloric acid,** and immediately introduced while wet into the hottest part of the flame. When the blue-green color is no longer seen, proceed to examine for **calcium.**

135. Red flames. Calcium is present when the red flame color, on evaporating the last portion of **hydrochloric acid,** appears siskin-green through the green glass. **Strontium** gives in this case a weak yellow. **Strontium** may be recognized by the purple or rose color which is seen through the blue glass, when the test, after moistening with **hydrochloric acid,** is evaporated to dryness in the flame.

Potassium gives a gray-blue mantle color and a rose-violet flame color. These colors appear reddish-violet through the blue glass, violet through a violet glass, and blue-green through a green glass. The test is to be moistened with **sulphuric acid,** and repeatedly exposed to the flame for a short time.

Chromic acid gives a dark brownish-red border color

and a rose-red mantle color. The dry test is to be moistened with concentrated sulphuric acid and held in the border. Chromic oxide gives no color, and is to be first oxidized to chromic acid by moistening with a solution of sodium hypochlorite and drying.

136. Yellow flames. Sodium gives an orange-yellow flame color, which in very large quantity appears blue, but in small quantity is invisible through the blue glass. Through the green glass the flame appears orange-yellow, even with the smallest quantity. This glass is particularly adapted to the recognition of sodium in all its compounds. The test should be moistened with sulphuric acid, dried, and held in the hottest point of the flame.

BUNSEN'S FLAME REACTIONS.

Almost all the reactions which can be performed by means of the blowpipe may be accomplished with greater ease and precision in the non-luminous flame of the gas-burner. This flame, moreover, possesses several peculiarities which render it available for reactions, by which the smallest traces of many substances occurring mixed together can be detected with certainty when the blow-pipe and even still more delicate methods fail. Only the principal reactions that can be obtained in this way are here given.

137. Bunsen's gas-lamp. This lamp, with non-luminous flame, is represented in Fig. 4, and must be made about three times as large as the drawing. It must be furnished with a cap for closing and opening the draught-holes, so as to be able to regulate the supply of air for every dimension of the flame. The conical chimney *d d d d* (Fig. 33) must also be made of such a size that the flame burns perfectly steady. Fig. 33 represents the

Fig. 33.

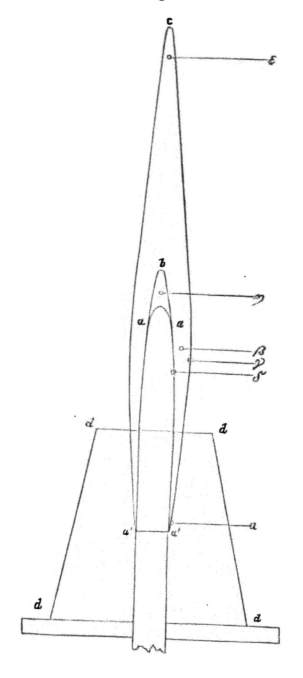

flame of its proper size. It is composed of the following three chief divisions:

A. **The dark cone,** *a a a a,* containing the cold unburnt gas mixed with about 62 per cent. of air.

B. **The flame mantle,** *a c a b,* formed of the burning coal-gas mixed with air.

C. **The luminous point,** *a b a,* not seen when the lamp is burning with the draught-holes open, but obtained of the size required for the reactions by closing these holes up to a certain point.

The following six points in the flame are used in the reactions:

1. **The base of the flame** lies at *a*; its temperature is comparatively very low, as here the burning gas is cooled by the upward current of cold air, and much heat is absorbed by the cold end of the metal tube. If mixtures of flame-coloring substances are held in this part of the flame, it is often possible to vaporize the most volatile constituent, and thus in the first few moments to obtain tints which cannot be observed at higher temperatures, because they then become masked by colors produced by the volatilization of the remaining substances.

2. **The zone of fusion** lies at *β*, somewhat above the first third of the flame in height, and midway between the inner and outer limits of the mantle at the point where the flame is thickest. This is the point in the flame which possesses the highest temperature, and it is therefore used in testing substances as regards their melting-point, their volatility, emissive power, as well as for all processes of fusion at high temperatures.

3. **The lower oxidizing flame** lies at *γ*, in the outer margin of the zone of fusion, and is especially suitable for the oxidation of substances dissolved in beads of fused salts.

4. **The upper oxidizing flame** at ε is form highest point of the non-luminous flame, and powerfully when the draught-holes of the lai open. This flame is suited for the oxidatio portions of substance, for roasting off volatil products, and generally for all those cases of which an excessively high temperature is not

5. **The lower reducing flame** lies at δ, on edge of the mantle next to the dark central the reducing gases at this point are mixed w atmospheric oxygen, many substances remain tered which become deoxidized on exposure t reducing flame. This point of the flame gives very valuable reactions which cannot be obtain blowpipe. It is especially available for reductio coal and in beads of fused salts.

6. **The upper reducing flame** is formed by th point η, produced over the dark zone when the of air is lessened by the gradual closing of the dr of the lamp. If this luminous point is made too l be found that a test-tube filled with cold water be ered with a film of lampblack: this never ough This flame contains no free oxygen, is rich in fin incandescent carbon, and hence it possesses far erful reducing powers than the lower reducing is especially available for reducing metals when i to collect them in the form of films.

METHOD OF EXAMINATION IN THE Ṿ PARTS OF THE FLAME.

A. Behavior of the Elements at High Tempe

138. This is one of the most important react can be employed for the detection and separat

stances. The possibility of producing, with the flame of the lamp alone, a temperature as high as or higher than

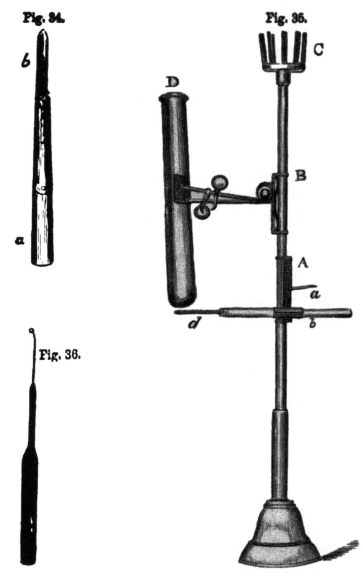

Fig. 34.

Fig. 35.

Fig. 36.

that of the blowpipe depends upon the fact that the radiating surface of the heated body be made as small as possible. The arrangement for bringing the substances into

I

the flame must therefore be on a very sn
platinum wire upon which the substance
scarcely exceed the thickness of a hor:
decimètre in length of the wire must 1
than 0.034 grm. It is impossible to o'
hereafter detailed if a thicker wire than tl
Substances which act upon platinum, or
adhere to the moistened surface of the m
the flame upon a thin thread of asbestos,
dred may be obtained from one splinter
These threads must not exceed in thickne
that of an ordinary lucifer-match.   De
stances are ground to the finest powder c
lamp-plate with the elastic blade (*a*) of
34), and drawn up on to a moistened stri
centimètre of filter-paper.   If the papei
being held with the platinum forceps, or,
two rings of fine platinum wire, the sam
coherent crust, which now may witho
heated in the flame.

If the substance require to be heated
a long period, the holder (Fig. 35) is 1
(*a*) is fastened to the carrier (A), so fixe
by a spring (as seen at B) that it can be 1
izontally and vertically.   The glass tube (
on this arm (*a*), and the fine platinum v
the tube thus held in the flame.   The sp
tos are stuck into the glass tube (*b*), whi
holder, and may then be moved with 1
The carrier (B) carries a spring-clamp f
tubes which have to be heated for a cons
a particular part of the flame.   The littl
contains nine upright supports to hold the

36) employed in the experiments. By means of these arrangements a particle of the substance under examination is brought into the flame, and its behavior in the coldest and hottest parts of the flame is ascertained, the substance being examined with a lens after each change of temperature. The following six different temperatures can be obtained in the flame, and these points may be judged of by observing the tints attained by the thin platinum wire:

1. Below a red heat.
2. Commencing red heat.
3. Red heat.
4. Commencing white heat.
5. White heat.
6. Strong white heat.

It is scarcely necessary to remark that these different temperatures must not be ascertained by the glow of the substances themselves, as the luminosity of different bodies depends not only upon the temperature, but also mainly upon their specific power of emission.

The following phenomena are observed when a sample of a substance is heated:

139. Emission of light. The emissive power of substances is ascertained by placing them on the platinum wire in the hottest part of the flame. The sample is of weak emissive power when it is less luminous than the platinum wire; of a mean emissive power when both appear about equally luminous; and of strong emissive power when the intensity of the light which it emits is greater than that from the platinum. Most solid bodies emit a white light, others—as, for instance, erbia—colored light.

Some bodies, such as many osmium, carbon, and molybdenum compounds, volatilize and separate out finely-

divided solid matter, which renders the flame luminous.
Gases and vapors always exhibit a smaller power of emission than fused substances, and these generally less than
solid bodies. The form of the substance under examination must always be noted, as the emissive power depends
upon the nature of the surface: thus, compact alumina,
obtained by slowly heating the hydrate, possesses only a
moderate emissive power, whereas the porous oxide prepared by quick ignition of the sulphate possesses a high
power of emission.

140. The melting point is determined by using the six
different temperatures already mentioned. At every in-
crease of temperature the bead is examined with the lens
to see whether the volume is decreased or increased, whether bubbles are given off on melting, whether on cooling the bead is transparent,
and what changes of color it undergoes during the
action of the heat or on afterward cooling.

Fig. 37.

141. The volatility is ascertained by allowing
equally heavy beads of the substance, placed on
a platinum wire, to evaporate in the zone of fusion, and observing the time, by means of a me-
tronome, which the bead takes to volatilize. The
point at which the whole of the substance is converted into vapor can be ascertained with great
accuracy, often to a fraction of a second, by the
sudden disappearance of the coloration of the
flame. The platinum wire upon which the substance is weighed is protected from the moistur
of the air by insertion in a tube (Fig. 37). If w
know the weight of the tube and wire, the right weigh
of substance can easily be attached, either by volatiliz
ing a portion or by fusing some more substance on t

the bead, and thus making it lighter or heavier. The experiments are best made with one centigramme of substance. The position in the flame where the highest constant temperature exists can be found by moving a fine platinum wire, fixed on a stand and bent at its point at a right angle, slowly about the zone of fusion, and noting the point where it glows most intensely. The beads to be volatilized are then most carefully brought into the flame at the same distance from the point of this wire. Care must also be taken that the dimensions of the flame do not undergo change from alterations in the pressure of the gas while the experiments are going on.*

142. Flame-coloration. Many substances which volatilize in the flame may be detected by the peculiar kinds of light which their glowing gases emit. These colorations appear in the upper oxidizing flame when the substance causing them is placed in the upper reducing flame. Mixtures of various flame-coloring substances are tested in the lowest and coldest part of the flame; and here it is often possible to obtain for a few moments the peculiar luminosity of the most volatile of the substances unaccompanied by that of the less volatile constituents.

B. Oxidation and Reduction of Substances.

In order to recognize substances by the phenomena exhibited in their oxidation and reduction, and to obtain them in a fit state for further examination, the following methods are employed:

143. Reduction in glass tubes is especially employed for the detection of Hg, or for the separation of S, Se, P, etc., when in combination with Na or Mg. A stock of

* For results of experiments by Hurtzig and Bunsen see *Flammen-reactionen*, by R. Bunsen, 1880.

very thin glass tubes is prepared, ~~~~~~~~~~
width and 3 centims. in length. ~~~~~~~~~
made out of one ordinary-sized test-tube, b
glass before the blowpipe, and then draw
the requisite size of tube is obtained.   Th
then cut up with a diamond into pieces 6
long, and each of these again divided into
lamp, and the closed ends neatly rounded.
having been finely powdered with the kni
34, *a*) on the porcelain plate (Fig. 38), i

Fig. 38.

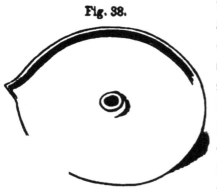

tube either by
a mixture of ca
or with sodiu:
sium. A pie
sium wire, a f
length, is for
pushed down
dered sample
the glass tube ;
carefully freed

and rolled out between the fingers to a st
which is then surrounded by the powder
The best form of carbon is the soot fr
which has been deposited upon the outsi
filled with cold water.   As soon as the sm
taining the perfectly dry sample has been
point of fusion of the glass, when general
inside the tube is noticed, it is allowed to c
placed upon the porcelain plate, covered by
per, and crushed to powder with the knife, f
of further examining the products of reduc

144. Reduction on splinters of charcoal. 1
metal can be obtained in small globules, o

A few tenths of a milligramme of the metal are generally sufficient to yield a solution with which all the characteristic precipitations can be accomplished, the reagents being contained in capillary glass threads, dropped into the solution by the milligramme, and the effect thus produced ascertained by examination with a lens. Iron, cobalt, and nickel, which do not fuse to globules on the splinter, are withdrawn from the agate mortar by means of the point of the magnetized blade (Fig. 134 *b*, washed with water, and dried high above the flame on the point of the knife. If the blade be then tightly drawn between the upper part of the thumb and the lower part of the first finger, and if the point of the blade be then approached to the metallic particles on the finger, they jump from the hand to the blade, forming a brush-like bundle, which can be conveniently examined by the lens, and by touching with a melted borax bead can be transferred in suitable quantities. The portion of metal remaining on the knife is rubbed on to a small piece of filter-paper, a drop of acid added, and the paper warmed over the flame so as to allow the metal to dissolve; this solution can then be further examined with various reagents.

c. Films upon Porcelain.

116. Those volatile elements which are reduced by carbon and hydrogen can be deposited from their compounds as films on porcelain, either in the elementary state or as oxides. Such films can be easily converted into iodides, sulphides, and other compounds, and thus may be made to serve as most valuable and characteristic tests. The films are composed in the centre of a thick layer, which on all sides gradually becomes thinner until the merest tinge is reached; it is therefore necessary to

distinguish between "thick" and "thin" parts of the films. Both exhibit in their variation of thickness all the tints of color characteristic of the substance under different circumstances of division. One-tenth up to one milligramme is in many cases sufficient for these reactions. Many surpass Marsh's arsenic test in delicacy and certainty, and approach in this respect the spectrum-analytical methods.

The following films can be obtained:

146. Metallic films are prepared by holding in one hand a particle of the substance on an asbestos thread in the upper reducing flame, which must not be too large, whilst with the other hand a glazed porcelain basin, one to two decimètres in diameter, filled with cold water, is held close above the asbestos thread in the upper reducing flame. The metals separate out as dead-black or brilliant-black films of varying thickness. Even Pb, Sn, Cd, and Zn yield in this way films of reduced metal, which by mere inspection cannot be distinguished from the soot separated out on the porcelain by a smoky flame. By means of a glass rod, these films can be touched with a drop of dilute HNO_3, containing about 20 per cent. of real acid; and the various degrees of solubility of the films serves as a distinguishing characteristic.

Cadmium,
Lead,
Zinc,
Indium. } The film dissolves quickly.

Bismuth,
Mercury,
Thallium. } The film dissolves slowly.

Antimony,
Arsenic,
Selenium,
Tellurium. $\Big\}$ The film is insoluble.

147. Oxide films are obtained by holding the porcelain basin filled with water in the *upper oxidising flame*, the rest of the operation being the same as in the production of the metallic films. If only a very small quantity of the sample can be employed, care must be taken to lessen the size of the flame, in order that the volatile products may not be spread over too large a surface of porcelain.

The film of oxide is examined as follows:

(*a*) The color of the thick and thin film is carefully observed.

(*β*) The reducing action or otherwise of a drop of stannous chloride is noted.

(*γ*) If no reduction occurs, NaHO is added to the stannous chloride until the precipitated hydrate redissolves, and then it is to be observed whether a reduction occurs.

(*δ*) A drop of perfectly neutral silver-nitrate is rubbed over the film with a glass rod, and a current of ammoniacal air is blown over the surface from a small wash-bottle containing ammonia solution, and having the mouth-tube dipping under the liquid and the exit-tube cut off close below the cork. If a precipitate is formed, the color is observed, and the solubility or alteration, if any, noticed, which occurs when the current of alkaline air is continued, or when a drop of ammonia liquor is added.

148. Iodide films are simply obtained from the oxide films by breathing on the latter upon the cold basin, which is then placed upon the wide-mouthed, well-stoppered glass (Fig. 39), containing fuming hydriodic acid and phosphorous acid derived from the gradual deliquescence of phosphoric tri-iodide. When the mixture no longer fumes, owing to absorption of moisture, it is easy to render it again fuming by adding a little phosphoric anhydride. Other films, often containing both iodides of a metal, and therefore frequently less regular in color and appearance, may be easily obtained by smoking the oxide film with a flame of alcohol containing iodine in solution, placed upon a bundle of asbestos threads, and held under the basin. If any iodine be condensed on the basin with the HI, it can easily be removed by gentle warming and blowing.

The examination of the film is conducted as follows:

(a) The solubility of the film is examined simply by breathing upon it when the basin is cooled; the color then either changes or entirely disappears, the film being dissolved in the moisture of the breath. If the basin be gently warmed, or if it be blown upon for some distance, the film again becomes visible by the evaporation of the moisture in the current of air.

Fig. 39.

(β) The ammonium compound of the iodide is formed by blowing ammoniacal air upon it, and noticing whether the color of the thick and thin films alters quickly, slowly, or not at all. The different colors reappear at once if the basin be held for a few moments over an open bottle containing fuming HCl.

(γ) The iodide films generally give the same reactions

TABLE OF VOLATILE ELEMENTS WHICH CAN

	METAL FILM.	OXIDE FILM.	OXIDE FILM WITH STANNOUS CHLORIDE.	OXIDE FILM WITH STANNOUS CHLORIDE AND SODIUM HYDRATE.	OXIDE FILM WITH SILVER-NITRATE AND AMMONIA.
Te	Black; thin part brown.	White.	Black.	Black.	Yellowish-white.
Se	Cherry-red; thin part brick-red.	White.	Brick-red.	Black.	White.
Sb	Black; thin part brown.	White.	White.	White.	Black; insoluble in ammonium hydrate.
As	Black; thin part brown.	White.	White.	White.	Lemon-yellow or reddish-brown; soluble in ammonium hydrate.
Bi	Black; thin part brown.	Yellowish-white.	White.	Black.	White.
Hg	Gray non-coherent thin film.				
Tl	Black; thin part brown.	White.	White.	White.	White.
Pb	Black; thin part brown.	Yellow-ochre color.	White.	White.	White.
Cd	Black; thin part brown.	Blackish-brown; thin part white.	White.	White.	White; in the thin parts turns bluish-black.
Zn	Black; thin part brown.	White	White.	White.	White.
Sn	Black; thin part brown.	Yellowish-white.	White.	White.	White.

BE REDUCED AS FILMS ON PORCELAIN.

IODIDE FILM.	IODIDE FILM WITH AMMONIUM HYDRATE.	SULPHIDE FILM.	SULPHIDE FILM WITH AMMONIUM SULPHIDE.	FLAME-COLORATION.	
Brown; disappears for a time on breathing.	Disappears altogether on blowing.	Black to blackish-brown.	Disappears for a time.	Upper reducing flame, pale-blue; upper oxidizing flame, green. No odor.	Elements whose reduction films are scarcely dissolved in dilute nitric acid.
Brown; does not wholly disappear on breathing.	Does not disappear on blowing.	Yellow to orange.	Orange, and then disappears for a time.	Cornflower-blue. Odor of rotten horse-radish.	
Orange-red to yellow; disappears on breathing.	Disappears altogether on blowing.	Orange.	Disappears for a time.	Upper reducing flame, pale-greenish. No odor.	
Orange-yellow; disappears for a time on breathing.	Disappears altogether on blowing.	Lemon-colored.	Disappears for a time.	Upper reducing flame, pale-blue. Odor of garlic.	
Bluish-brown; thin parts pink; disappear for a time on breathing.	Pink to orange; chestnut-colored when blowing.	Burnt-umber-color to coffee-colored.	Does not disappear.	Bluish; not characteristic.	Elements whose reduction films are with difficulty dissolved in dilute nitric acid.
Carmine-colored and lemon-yellow; does not disappear on breathing.	Disappears for a time on blowing.	Black.	Does not disappear.		
Lemon-yellow; does not disappear on breathing.	Does not disappear on blowing.	Black; thin parts bluish-gray.	Does not disappear.	Light grass-green.	
Orange-yellow to lemon-color; does not disappear on breathing.	Disappears for a time on blowing.	Brownish-red to black.	Does not disappear.	Light-blue.	Elements whose reduction films are instantly dissolved in dilute nitric acid.
White.	White.	Lemon-colored.	Does not disappear.		
White.	White.	White.	Does not disappear.		
Yellowish-white.	Yellowish-white.	White.	Does not disappear.	Intense indigo-blue.	

as the oxide films with silver nitrate and ammonia, with stannous chloride, and with caustic soda.

149. The sulphide film is most easily obtained from the iodide film by blowing upon it a current of air saturated with ammonium sulphide, and removing the excess of sulphide by gently warming the porcelain. It is advisable to breathe on the film from time to time whilst the current of sulphuretted air is being blown on the basin. The experiments to be made with this film are:

(*a*) The solubility or otherwise in water is ascertained by breathing on it, or by addition of a drop of water. The sulphides often possess the same color as the iodide films; they may, however, generally be distinguished by their insolubility on breathing.

(*β*) The solubility of the sulphide in ammonium sulphide is ascertained by blowing or dropping.

150. Films on test-tubes. Under certain circumstances it is advisable not to collect the film on porcelain, but upon the outside of a large test-tube (**Fig. 35, D**); this method is especially used when it is needed to collect larger quantities of the reduction film for the purposes of further examination. The fine asbestos thread with the sample of substance is held on the glass tube (*b*) before the lamp, so that it is placed at the height of the middle of the upper reducing flame, and the test-tube fixed so that the lowest point is just above the end of the asbestos thread. If the lamp be now pushed under the test-tube, the substance and the asbestos thread are in the reducing flame. By repeating this operation, the film can be obtained of any desired thickness; some pieces of marble are in this case placed in the test-tube, to prevent the water from being thrown out of the tube by percussive boiling.

THE REACTIONS OF THE ELEMENTS.

The elements, which can easily be recognized by their flame reactions, are arranged in the following groups and sub-groups according to their behavior in the reducing and oxidizing flames:

A. Elements whose compounds are reducible to metal and form a film on porcelain (see page 143):

1. Films scarcely soluble in cold dilute nitric acid (containing about 20 per cent. of acid)—*tellurium, selenium, antimony, arsenic* (pages 143–146).

2. Films slowly and difficultly soluble in cold dilute nitric acid—*bismuth, mercury, thallium* (pages 146–148).

3. Films instantly soluble in cold dilute nitric acid—*lead, cadmium, zinc, indium* (pages 148, 149).

B. Elements whose compounds are reduced to metal, but form no film:

1. Not fusible to a metallic bead after reduction.
   a. Magnetic—*iron, nickel, cobalt* (pages 150, 151).
   b. Non-magnetic—*palladium, platinum, rhodium, iridium, osmium* (pages 151, 153).

2. Fusible to metallic beads—*gold, silver, copper, tin* (pages 153, 154).

C. Elements most easily separated and recognized as compounds—*Molybdenum, tungsten, titanium, lantalum, niobium, chromium, vanadium, manganese, uranium, silicon, phosphorus, sulphur* (pages 155–162).

A. ELEMENTS WHOSE COMPOUNDS ARE REDUCIBLE TO METAL, FORMING A FILM UPON PORCELAIN.

151. Tellurium compounds. *Flame-coloration*, in upper

reducing flame, pale-blue, whilst the oxidizing flame above appears green.

Volatilisation, unaccompanied by any odor.

Reduction film, black, with dark-brown coating, dull or brilliant ; heated with concentrated sulphuric acid, gives a carmine-red solution.

Oxide film, white, scarcely or not at all visible ; stannous chloride colors it black, by reason of separated tellurium ; silver nitrate, after ammonia has been blown upon it, yellowish-white.

Iodide film, dark-brown, with brown coating ; disappears momentarily when breathed upon, but not when slightly warmed ; reappears on exposure to HCl ; blackened by $SnCl_2$.

Sulphide film, dark-brown to black ; does not disappear when breathed upon ; dissolves in NH_4HS blown upon it, and reappears upon warming or if blown upon with air.

With soda on charcoal splinter gives a sodium telride, which, when moistened upon a silver coin, oduces a black spot ; and if the specimen contains uch tellurium, with HCl, diffuses an odor of hyogen telluride with the separation of black tellurum.

152. Selenium compounds. *Flame-coloration,* pure azure-u.

Volatilizes, burning with the odor of selenium.

Reduction films, brick-red to cherry-red ; at one time ll, at another brilliant ; gives, when heated with connitrated H_2SO_4, a dirty-green solution.

Oxide film, white ; brick-red from separated selenium n $SnCl_2$ is dropped upon it ; the old color darkened aHO ; with $AgNO_3$, the oxide film gives a white, scarcely visible, coloration, which disappears when ammonia is blown upon it.

Iodide film, brown; contains some reduced selenium, and therefore cannot be made to disappear completely, either by breathing upon it or by blowing ammonia upon it.

Sulphide film, yellow to orange-red; insoluble in water, soluble in NH_4HS.

With soda on charcoal splinter gives sodium selenide, which produces, with a drop of water, a black spot upon a silver coin, and moistened with HCl, if the quantity is not too small, gives the odor of hydrogen selenide, with separation of red selenium.

153. Antimony compounds. *Flame-coloration*, by treatment in the upper reducing flame, pale-green, unaccompanied by any smell.

Reduction film, black; sometimes dead, sometimes bright.

Oxide film, white; moistened with a perfectly neutral solution of $AgNO_3$, and then blown on by ammoniacal air it gives a black spot which does not disappear in NH_4HO. If the film be first placed over bromine vapor the reaction cannot be obtained, owing to the oxidation of Sb_2O_3 into Sb_2O_5. It is unaltered by $SnCl_2$, either with or without NaHO.

Iodide film, orange-red, disappearing by breathing, and reappearing by blowing or warming; blown on with ammoniacal air it disappears, but does not return. Generally it gives the same reactions as the oxide.

Sulphide film, orange-red. The film is difficult to blow away with NH_4HS; returns on blowing with air; insoluble in water.

With soda on charcoal splinter gives no black stain on silver, but yields a white, brittle, metallic bead.

154. Arsenic compounds. *Flame-coloration*, in upper reducing flame, pale-blue, giving the well-known arsenical smell.

Reduction film, black, dead, or brilliant; thin film brown.

Oxide film, white; touched with a perfectly neutral solution of $AgNO_3$, and then blown with ammoniacal air it gives a canary-yellow precipitate, soluble in NH_4HO. Together with this yellow precipitate, a brick-red one of silver arsenite occurs when the film has previously been treated with bromine vapor. $SnCl_2$, with and without soda, produces no change.

Iodide film is deep-yellow; disappears on breathing, but returns on drying; disappears in ammoniacal air, and does not return; reappears unaltered after the action of HCl.

Sulphide film, lemon-yellow; disappears easily on blowing with NH_4HS, and reappears on warming or blowing; insoluble in H_2O, and does not disappear by blowing upon it.

Reduction on charcoal splinter yields no metallic bead.

155. Bismuth compounds. *Reduction film*, black, dead, or brilliant; thin portion of film, brownish-black.

Oxide film, light-yellow; unaltered by $AgNO_3$, with or without ammonia; gives no reaction with $SnCl_2$, but yields black precipitate of $BiHO_2$ on addition of NaHO.

Iodide film is very characteristic, and remarkable for the number of tints which it assumes. The thick part is of a brown or blackish-brown color, with a shade of lavender-blue; the thin film varies from flesh-color to light-pink; it easily disappears on breathing, and appears again on blowing. In a stream of ammoniacal air it passes from pink to orange, and on blowing or warming it again, at-tains a chestnut-brown color; it resembles the oxide film in its behavior with $SnCl_2$ and NaHO.

Sulphide film is of a burnt-umber color; the thin parts are of a lighter coffee-brown color; does not disappear on blowing, and is not soluble in NH_4HS.

On charcoal splinter with soda the bismuth compounds are reduced to a metallic bead, yielding, when rubbed in the mortar, bright, shining, yellowish splinters of metal soluble in HNO_3. The solution gives, with $SnCl_2$ and NaHO, black $BiHO_2$.

156. Mercury compounds. *Metallic film* is mouse-gray, non-coherent, and spreads over the whole basin. To obtain small traces of Hg in the reduced state, the sample is mixed with soda and KNO_3 and filled into a thin test-tube five to six millims. wide and ten to twenty millims. long. This is held by a platinum wire in the flame, whilst the bottom of the basin, filled with cold water, is placed close above the open end of the tube. If the quantity of Hg is considerable, it collects in the form of globules, which can be seen with a lens, and which can be collected into larger drops by wiping the basin with a piece of moistened filter-paper.

Iodide film is obtained by breathing on the metallic film, and then placing it over the vessel (Fig. 39, page 139) containing moist Br. It first becomes black, and then disappears, but not until after some time; HgBr is formed. If the basin be now placed aboye the vessel of fuming HI, a very characteristic carmine-colored film of Hg_2I is produced; this is often accompanied by HgI, but neither of these disappear when breathed upon or when blown upon with ammoniacal air.

Sulphide film, black; not altered by breathing or by blowing with NH_4HS.

157. Thallium compounds. Since t
this element can be recognized by in
scope, it will seldom be detected in a

Flame coloration, bright-grass-gree
Metallic film, black, with coffee-br
Oxide film, colorless; unchanged
also with $AgNO_3$ with or without NH

Iodide film, lemon-yellow; insolub
Sulphide film, obtained from the o
livid coating; insoluble in NH_4HS.

On charcoal with soda, reducible
grain.

158. Lead compounds. *Flame-colo*
Reduction film, black, dead, or bri
Oxide film, bright yellow-ochre col
ride gives no reaction even on additio
does not produce any reaction, either
of NH_4HO.

Iodide film, orange- to lemon-ye
breathing or on moistening; disappe
ammoniacal air, and again appears
Sulphide film, brownish-red to bla
moistening with NH_4HS it remains

On charcoal splinter with soda giv
ductile metallic bead, which is slowly
uble in HNO_3, yielding a white, easil
soluble in H_2O, and precipitated as
addition of HSO_4 from a capillary

159. Cadmium compounds. *Metall*
thin parts, brown.

Oxide film, brownish-black, shadin
to a white invisible film of oxide, w
by $SnCl_2$, alone or with soda; $AgN($

ish-blue coloration of reduced metal, which is very characteristic, and does not disappear on addition of NH$_4$HO.

Iodide film, white; no change produced by NH$_4$HO.

Sulphide film, lemon-yellow; insoluble in NH$_4$HO.

Reduction on charcoal splinter with soda. The metal,
owing to its volatility, can be obtained only with difficulty as a silver-white, ductile bead.

160. Zinc compounds. *Reduction film,* black; in the
thin parts, brown.

Oxide film, white, and therefore invisible. To test it,
a square centimetre of filter-paper, moistened with HNO$_3$,
is rubbed over the surface, and then rolled up on two
rings on fine platinum wire, three millimètres in diameter, and burnt. If the paper is burnt in the upper oxidizing flame at as low a temperature as possible, the ash
forms a small solid mass about a square millimètre in
area, which can be ignited without fusion, and becomes
yellow on gently heating, and appearing white on cooling. If this be moistened with a few milligrammes of
very dilute cobalt solution and ignited, it appears of a
beautiful green color on cooling; the same reaction can
be effected with the metallic film.

Iodide film, white; not clearly recognizable either
alone or after ammonia has been blown upon it.

Sulphide film, also white, and not easily recognized
either alone or when moistened with NH$_4$HS.

Reduction on charcoal splinter does not proceed on account of volatility of the zinc.

161. Indium compounds. Detected with most ease and
certainty with the spectroscope.

Flame-coloration, intense; pure indigo-blue.

Metallic film, black, with brown coating; at one time
dull, at another brilliant. Disappears instantly with HnO$_3$.

Oxide film, yellowish-white; scarcely vi
reactions with SnCl, and AgNO₃ solution

Iodide film, also yellowish, nearly white
weak, with and without ammonia.

Sulphide film, also yellowish; nearly v
visible. Unchanged by NH₄HS.

Reduction on charcoal splinter with so
with difficulty, and affords silver-white, di
slowly soluble in HCl.

B. ELEMENTS WHOSE COMPOUNDS A TO THE METALLIC STATE, BUT FOR

1. *METALS NOT FUSED TO A BEA REDUCTION.*

a. Magnetic metals.

162. Iron compounds. *Reduction on ch*
gives no metallic bead or ductile lustrous
finely-divided metal forms a black brush
the magnetized knife-blade; this, when
paper and dissolved in a drop of aqua
yellow spot when warmed over the flame
moistened with potassium ferrocyanide, gi
oration of *Prussian blue.* The yellow s
with NaHO, and then held for a few mon
sel with bromine vapor, gives, on a secon
soda, no coloration of a higher oxide.

Borax bead. In the oxidizing flame, wh
to brownish-red; when cold, yellow to bro
reducing flame, bottle-green.

163. Nickel compounds. *Reduction on the*
ter. On pulverizing the charcoal, white, lu
metallic particles are obtained, forming a

magnetized blade. The metal, dissolved in HNO_3 on paper, gives a green solution, which, on moistening with soda, exposure to bromine vapor, and second addition of soda, give a brownish-black spot of Ni_2O_3. The ash of the paper, from which the soda has been washed out, can be used for the borax-bead test.

Borax bead. Oxidizing flame, grayish-brown or dirty-violet. Upper reducing flame, gray, from reduced Ni, which often collects to a spongy mass of metal, rendering the bead colorless.

164. Cobalt compounds. *Reduction on charcoal splinter.* By pulverizing the charcoal, white, ductile, lustrous, metallic particles are obtained, which form a brush on the magnetic blade. The metal, rubbed off on to paper, gives a red solution when moistened with HNO_3; this yields a green color on addition of HCl and drying, which disappears again on moistening. The paper, moistened with soda, brought into bromine vapor, and again moistened with soda, yields a brownish-black spot of Co_2O_3. This reaction is plainly seen with a few tenths of a milligramme of metal. The paper can also be used, after washing out the soda and burning, for the coloration of the borax bead.

Borax bead. Deep-blue bead in the oxidizing flame, which does not change in the lower reducing flame. When treated for a considerable time alone, or, better still, with ammonium platin-chloride, in the most energetic upper reducing flame, is completely decolorized, but only after long treatment, with the separation of cobalt or platin-cobalt.

b. Non-magnetic metals.

165. Palladium compounds. These are reduced upon fine *platinum wire, with soda,* in the upper oxidizing

flame, to a gray mass, similar to platinum sp
when rubbed in an agate mortar, gives shining
tallic scales. The scales, rinsed and dried on
glass plate, dissolve in HNO_3, with reddish-

If a small drop of a solution of mercuri
added to the liquid, a white flocculent prec
tained, which dissolves in NH_4HO when d
it. After evaporation and boiling with aq
liquid, evaporated to a small drop, gives a
yellow, crystalline precipitate of palladiu
chloride.

Solution of palladium is colored blue,
brown by $SnCl_2$, according to the amount

166. Platinum compounds. These give,
upon platinum wire with soda in the upp
flame, also a gray spongy mass, which, by
an agate mortar, is converted into shining,
ductile, metallic scales. These are insolub
HNO_3, or HCl alone, but with aqua regia gi
yellow solution if the platinum is pure; if
rhodium, iridium, or palladium, they give
yellow solution. When solution of mercuri
added to the solution and ammonia blown
flocculent white precipitate is formed, but is
bright-yellow crystalline precipitate of amm
chloride.

$SnCl_2$ colors solutions of platinum yellow

167. Iridium compounds. These, ignited
oxidizing flame with soda, are likewise redu
which, when rubbed in an agate mortar, g
powder without lustre, and not in the least
tile. This is insoluble in nitric acid, hydro
and aqua regia.

168. Rhodium compounds. These are only distinguished from the iridium compounds by the fact that the metallic powder, insoluble in aqua regia, when fused with acid potassium sulphate, is partially oxidized, and affords a rose-red solution.

169. Osmium compounds. These give in the oxidizing flame volatile osmium tetroxide of a pungent odor, similar to chlorine, and which irritates the eyes.

2. *METALS FUSED TO A BEAD AFTER REDUCTION.*

170. Gold compounds. If only traces of gold are present, mixed with a considerable quantity of gangue, it can only be concentrated and detected according to the old processes for detecting gold. Otherwise even a few tenths of a milligramme can be recognized by reduction with soda on a charcoal splinter. The yellow, shining, ductile, metallic grain obtained in this way can be reduced to spangles having the lustre of gold by rubbing in an agate mortar. These are insoluble in HCl or HNO_3, but give rather readily, with aqua regia, a bright-yellow solution. If this is soaked up into a piece of filter-paper and touched with $SnCl_2$, purple of Cassius is formed. What remains upon the glass is colored brown by a solution of $FeSO_4$, by reason of separated gold, whilst the liquid appears blue by transmitted light.

171. Silver compounds. If silver occurs only in traces in slags or complex ores, it can only be detected by the well-known method of cupellation. If, however, the silver compound is not mixed with a very large amount of foreign matter, it can be detected in very minute quantities by reduction with soda on the charcoal splinter. The

anide. Instead of acting upon a metal in a curved glass, it may be dissolved by moistening paper upon which it is placed with HNO_3.

With borax on platinum wire. Blue bead, not altered to cuprous oxide when heated in the lower reducing flame alone, but on addition of very little tin oxide, forms a reddish-brown bead. If this bead be frequently oxidized and reduced in the flame, a ruby-red transparent bead is obtained; this occurs most readily when the bead is allowed to oxidize very slowly.

173. **Tin compounds.** *On the charcoal splinter* the Sn compounds are easily reduced to white, lustrous, ductile, metallic beads. The flattened particles, transferred to the curved glass, slowly dissolve in HCl; and the solution, when absorbed by paper, gives a red precipitate with selenious, and a black precipitate with tellurous acid, dissolved in HCl. If to the solution a trace of bismuth-nitrate be added, an excess of soda gives a black precipitate of Bi_2O_3. The metal, acted on by HNO_3, yields a white powder of insoluble metastannic acid.

A borax bead, containing enough CuO to render it

faintly blue, serves as a delicate test to ascertain with certainty the presence of a trace of a Sn compound, as the bead, placed in the lower reducing flame, turns reddish-brown or forms a clear ruby-red glass.

C. ELEMENTS MOST EASILY DETECTED BY THE REACTIONS OF THEIR COMPOUNDS.

a. Metallic substances.

174. Molybdenum compounds. *On a charcoal splinter with soda* are reduced, with great difficulty, to a gray powder. In the same way some compounds give in the upper reducing flame a film on porcelain, which it is difficult to obtain. Molybdenum is best recognized as follows:

a. The sample is finely pulverized with the knife on the porcelain plate, is mixed on the hand with soda, obtains a pasty state by fusion. The mixture is then transferred to a spiral of fine platinum wire and fused in the flame. The liquid fused mass is then knocked off the wire and allowed to fall upon the plate, when it is digested with two or three drops of water, and the clear liquid above the sediment is soaked up into three or four strips of filter-paper, not too fine, several millimètres broad. One of these strips, moistened with HCl, does not change color, but with a drop of potassium ferrocyanide is changed to reddish-brown.

If one of these strips is gradually moistened with a few milligrammes of $SnCl_2$, it either becomes blue in the cold or upon warming; if it becomes yellow or yellowish-brown, more of the solution of the test specimen must be added by means of a capillary pipette, in order to cause the blue color to appear.

in the lower reducing flame, a gray mass is formed, which dissolves on heating in HCl, yielding a pale amethystine-colored solution.

177, 178. Tantalum and niobium compounds. These manifest the same reactions as titanium.

179. Chromium compounds. In platinum spiral with soda, the compounds when fluxed, with the repeated addition of potassium nitrate, give a bright-yellow mass, which, when knocked off on to the porcelain plate and crushed, give a bright-yellow solution. If this solution is decanted from the residue, and acetic acid added, it becomes yellowish-red, and gives ·with lead salts, when it is soaked up by strips of filter-paper, a yellow precipitate; with solutions of salts of mercury oxide, a red one; and with $AgNO_3$, a reddish-brown one.

With NH_4S, also by evaporation with aqua regia upon the porcelain plate, the solution becomes green; likewise with $SnCl_2$. The *borax bead* becomes emerald-green in the oxidizing flame, and does not change this color in the reducing flame.

180. Vanadium compounds. Treated with soda and *potassium nitrate in a platinum spiral*, yield a bright-yellow mass, the solution of which, on addition of $AgNO_3$ and acetic acid, yields a yellow precipitate. The fused mass, when evaporated with aqua regia, gives a yellow instead of a green solution, which becomes blue on addition of $SnCl_2$. If much vanadium is present, the solution gives a yellowish-brown solution or precipitate on addition of concentrated cold HCl. In the *borax bead* these compounds give a yellowish-green color in the oxidizing flame; in the reducing flame, a green color.

181. Manganese compounds. *Borax bead.* Amethyst in the oxidizing flame; colorless in the reducing flame.

14

With sodium carbonate on platinum wire, a bead is formed green after cooling, especially easily after addition of potassium nitrate. Water extracts a green solution from it, which becomes red after the addition of acetic acid, and then, often with the separation of brown flakes, becomes colorless.

182. **Uranim compounds** give a yellow bead in the oxidizing flame, which becomes green in the reducing flame, especially on addition of $SnCl_2$. These colors closely resemble those of the iron compounds, but may easily be distinguished, at least if no other coloring metallic oxide is present, by the fact that the uranium bead, when incandescent, emits a bluish-green light, analogous to that which the uranium compounds exhibit when fluorescing. Beads of lead oxide, stannic acid, and a few other substances exhibit a similar phenomenon when incandescent, but they do not yield, like uranium compounds, a colored bead on cooling.

Heated gently on the *platinum spiral with HKSO₄*, the insoluble uranium compounds can be decomposed. The melted mass is powdered with a few particles of crystallized sodium carbonate, and the moistened mass is absorbed by filtering paper. A brown spot is formed by the addition of a drop of potassium ferrocyanide to the moistened paper.

the blood-red coloration in a bead with iron sesquioxide, produced by these acids. The fluxed mass, if water and acetic acid are carefully added, and then evaporated on the porcelain plate, separates gelatinous hydrated silica. Fine splinters of silicate give, upon fusing in the bead of salt of phosphorus, a gelatinous skeleton of silica, floating in the fused or cooled bead.

184. Phosphorus compounds. These may easily be detected in its compounds, even when they are mixed with large quantities of other substances, as follows:

The sample, having been ignited, is rubbed fine on the porcelain plate (see **Fig. 38**), and is then introduced into a small glass tube of the thickness of a straw; into this tube, which is closed at the bottom, a piece of magnesium wire, about one-fourth of an inch in length, is placed so that it is covered by the powder. On heating the tube, magnesium phosphide is formed with incandescence. The black contents of the tube powdered on the plate give, on moistening with H_2O, the highly-characteristic smell of H_3P. A piece of sodium may be substituted for the magnesium wire.

If it has been ascertained that the sample does not yield any film on porcelain in the upper oxidizing flame, the phosphates may be recognized by heating on platinum with borax and a thin piece of iron wire in the hottest part of the reducing flame, when a bright molten bead of iron phosphide is obtained, which can be extracted with the magnetized knife on crushing the bead under paper.

185. Sulphur compounds. *On a charcoal splinter with soda*, in the lower reducing flame, they give a fused mass, which, moistened upon a piece of silver, blackens it. Since selenium and tellurium produce the same reaction, the absence of these substances must be determined by the absence of a tellurium or selenium spot upon porcelain.

When only metallic sulphides are to be considered, and not sulphates, it will answer simply to heat the test specimen in the flame to detect sulphur by the odor.

EXAMPLES SHOWING THE APPLICATION OF THE FOREGOING METHODS.

186. *a*. **A mixture of sulphide of arsenic, sulphide of antimony, and sulphide of tin.** If in a mixture of these three sulphides, containing only traces of Sb and Sn, they are separated according to the ordinary rules of qualitative analysis, by dissolving in alkaline sulphides and reprecipitation with acids, the detection of these two metals by the regular tests is extremely uncertain and troublesome. According to the following method the detection of these metals is rendered easy and certain when the proportion of Sn is only a few thousandths, and that of the Sb only a few hundredths, of the total weight of the mixture.

Three decigrammes of the sulphides are roasted on a curved piece of glass * small enough to be altogether surrounded by the flame, and the residue, weighing only a few milligrammes, is scraped together with the knife. The moistened mass is then collected on the end of a thread of asbestos and a strong metallic film obtained on the test-tube.   In order to prevent the deposition of any carbon with the metals, which would act injuriously in the subsequent operations, the upper reducing flame is made so small that the luminous point is only just visible.   The film is next dissolved in a drop or two of HNO_3 in the curved rim (Fig. 38, page 134), and the solution evaporated below its boiling-point by gently

---

* Pieces of a thin chemical flask are also best for use in this case.

warming and blowing, so as to obtain the solid residue in as small a space as possible. A drop of neutral silver solution is now brought on to the residue at the moment when it becomes solid; and on blowing with ammoniacal air a characteristic black stain is formed, whilst the reaction of As is also generally noticed.

In order to detect Sn, a few scarcely visible particles of the roasted sulphides are fused on to a borax bead which has been very slightly tinted with cupric oxide. If the bead is now brought into the lower reducing flame, it becomes a ruby-red color from reduced cuprous oxide. If the oxide be present in too large a quantity, the bead can be obtained transparent by the process described under the reactions of the copper compounds. This reaction can only be obtained in the lower reducing flame of the non-luminous gas-lamp, as in the ordinary blowpipe flame the cupric oxide is reduced to cuprous oxide without the presence of tin-salt.

187. *b.* **Black tellurium, containing tellurium, selenium, antimony, gold, lead, and sulphur,** After the sulphur has been detected by the smell by roasting, the metallic film is obtained on a test-tube, which is then placed inside a wider and shorter tube containing a few drops of *concentrated* sulphuric acid, so that the metallic film is surrounded by the acid. If the temperature be now gradually raised the presence of tellurium is at once ascertained by the formation of a bright carmine color. If the temperature be still further raised, the tellurium oxidizes and the olive-green color of selenium becomes visible; the cooled solution, on dilution with water, then no longer exhibits the black precipitate of tellurium, but is colored yellowish-red with the selenium. If this is present in small traces only, it can be best detected by looking

14*   L

down the length of the ~~text tube~~
paper. As common com~~mercial~~
traces of selenium, it is well to make a (
first. The antimony is detected as in pre(
To detect the lead and gold, a sample is
charcoal splinter, the beads of the alloy
a curved glass, and the flattened and dri(
ticles treated with rather strong HNO_3
thing dissolves. The acid is then evap
the soluble portion of the residue dissol\
two of water. The solution is brought
curved glass by means of a capillary |
characteristic precipitate of lead sulphat(
H_2SO_4. The gold left undissolved as
der is completely washed by frequent a(
and removal of the same with the capill
portion of the dried particles of gold fus(
splinter with soda yields in the mortar
golden particles, which may be dissolve(
and tested with $SnCl_2$. A centigramme I
sufficient in experienced hands for all the

SPECTRUM ANALYSIS.

188. The second method—that of *sp*
discovered by Kirchhoff and Bunsen—
ting the rays of the colored flame, after
a narrow slit, traverse a prism, and in ob(
trum thus produced by means of a teles(
the metals which give color to the flan
peculiar spectrum, formed in some case
barium, of many contiguous colored li■
two more distant lines of different color, I
ium; or, again, of a single line, as in cas(

thallium. These spectra are characteristic in two respects
—viz. 1, in the definite color of the spectrum lines; and
2, in the invariable relative position they occupy.

The last-named fact enables us to detect, without difficulty in most cases, all the spectrum-giving ingredients of
a mixture. Thus, when potassium, sodium, and lithium-
salts are brought together into the *spectroscope*, the lines
characteristic of each metal appear in the utmost purity at
one view. Very minute traces of some elements do not
however, exhibit their spectra in presence of large quantities of other substances.

The methods of obtaining the spectra of the elements
or their compounds vary according to their volatility.
The instrument or spectroscope used varies according to
the degree of accuracy which the observations require.
The direct-vision pocket spectroscope may be used for
most blowpipe investigations. Although it has no scale,
the relative positions of the various lines can be easily
determined from the table.

A much more powerful and perfect spectroscope, ex-
clusive of the source of light, is composed of an adjust-
able slit, a contrivance (collimating lens) for rendering
the rays parallel that have passed through the slit, and a
prism. All light except that under examination must be
excluded from the prism, and therefore the slit, prism,
and lenses are enclosed in a tube, or, if the prism be too
large, the latter is fitted with a separate cover. As the
spectrum on emerging from the prism is but little longer
than the width of the slit, and only becomes of some
length as the distance from the prism increases, a magnifying-glass is introduced, in order that the eye, though
at but a small distance from the prism, may see the spectrum of a sufficiently large size, and the spectrum, there-

fore, is not observed with the naked ey
medium of a telescope of moderate ｐ
scope is also necessary for enabling t
the whole of the light passing from th
through the prisms.

In order to know the exact po͏
tube is added, in the end of whi͏
tographed on glass, and which is so͏
server sees the scale and spectrum at th

A small glass prism may also bè attac
the tube through which the colored fla
that the spectra of two flames may be e
pared at the same time.

Not only the number of the spectru
stance, but also the degree of their inte
of careful attention.   As the brillianc
creases with the temperature, so, as ａ
lines which are particularly prominen
of temperature that are the first to ap
perature.   These are best suited for th
substance, and are therefore called *c*
Such'lines, according to their brightne
in each substance by the letters of the ｔ
β, γ, δ, etc., being affixed to the symbc

The table at the beginning of the
spectra of some of the more commo
easily recognized elements.

189. Potassium.   All the volatile coﾍ
sium, when placed in the flame, give ａ
continuous spectrum, which consists o
one line K_a, situated in the outermost
line, $K\beta$, situated far in the violet rays
end of the spectrum.

When vapor of potassium is heated in the electric spark, several other lines make their appearance. Before testing potassium silicates they should be ignited with sodium carbonate, as it does not interfere with the reaction. Orthoclase, sanidin, and adularia may be easily distinguished from albite, oligoclase, anorthite, and labradorite. If only a trace of potash is present, the silicate should be heated with ammonium fluoride in a platinum dish, and the residue placed in a flame with a platinum wire.

190. Sodium. The yellow line, Naa, is the only one which appears in the sodium spectrum as seen in the flame with the ordinary spectroscope. With powerful instruments it is seen to be double. The line a is remarkable for its definite form and brightness, and is produced by all the natural compounds of sodium.

191. Lithium. The salts of this metal give a bright line in the red, Lia, and another much less distinct in the orange, Liβ. With high heat and strong prisms a blue line also appears. All lithium compounds give the reaction, and often only require to be held in the flame, as lepidolite, petalite, etc. If the amount is very small in the silicate, it should be digested and evaporated with ammonium fluoride, a little sulphuric acid added, again evaporated, and the residue treated with alcohol. The solution in alcohol is evaporated to dryness, the mass again treated with alcohol, and the liquid dried in a glass capsule. The crust formed can be placed in the flame with a platinum wire.

192. Strontium. The spectra of the alkaline earths are more complex than those of the alkalies. Strontium gives eight very distinct lines—six red, one orange, one blue. The orange line, Sra, close to the sodium line, the two red lines, Srβ and γ, and the blue line, Srδ, are the most

important. The chloride ~~gives the line~~
the non-volatile compounds ~~give nothing~~
must be reduced to sulphide by ~~holding in~~
reducing flame, and the silicates must be
dium carbonate, powdered, and washed
decantation. The insoluble carbonate t
moistened with hydrochloric acid, and v
distinct reaction. The strontium lines c
with the indications of the alkalies.

193. Calcium. The spectrum of this
distinguished from all the foregoing by
Caβ, and by the orange, Caa. A feeble l
in the violet with a powerful instrument,
with increased heat. The chloride give
tion. The non-volatile compounds must
by hydrochloric acid and the silicates by
oxide. The composition of calcareous rocl
is easily found.

194. Barium. The complicated spectru
is distinguished by its green bands, of w
are the most important. The haloid salts
mon natural compounds are recognized b
in the flame. The silicates must first be l
drochloric acid or fused with sodium carl
dissolved in acid. If barium and strontiur
quantities with calcium, the carbonates ob
are dissolved in nitric acid and the driec
with alcohol. The residue contains or
strontium, which can generally be detec

195. Rubidium. The continuous spec
extended as that of potassium. The a l
brilliant and best suited for the recognitic
The lines δ and γ are less intense, but still

istlc. These lines, and even others, appear with not only the volatile chloride, nitrate, etc., but even with the silicates.

196. Cæsium. The spectrum is characterized by two lines, Csα and Csβ, both brilliant and well defined. The absence of any line in the red distinguishes this from the two previous spectra. With an intense light, yellow and green lines may be seen in the continuous parts of the spectrum. Rubidium and cæsium can be easily separated from lithium and sodium by means of platinic chloride, but potassium ·is precipitated with them, and must be removed by repeated boiling with water before the presence of these two elements can be proved by spectrum analysis.

197. Thallium. The compounds of this metal give a spectrum with a single intense green line, Tlα, which almost coincides with the δ line of barium. Mere traces of thallium in pyrites may be easily detected by simply heating them in the edge of the flame.

198. Indium. The spectrum is characterized by two lines, Inα in the indigo, and Inβ in the violet. The former is far the most intense, and sufficient for the detection of the metal. To show its presence in sphalerite (zinc blende), the mineral is roasted, decomposed by hydrochloric acid, and the solution diluted and saturated with ammonium hydrate. The precipitate containing the indium oxide is dried, and a small portion of it moistened with hydrochloric acid and placed in the flame with platinum wire. The presence of indium will be indicated by the blue line α.

199. It is not only those bodies which have the power of giving color to the flame which yield characteristic spectra, for this property belongs to all elementary sub-

stances, whether metal or ▓▓▓▓▓▓▓▓
and it is always noticed ▓▓▓▓▓▓▓
the point at which its vapor ▓▓▓▓▓▓
each element emits the ▓▓▓▓▓▓▓
alone, and the characteristic ▓▓▓▓▓▓
ent when its spectrum is observed. ▓▓▓
a much higher temperature than the ▓▓
order that their vapors should become
they may be easily heated to the requis
by means of the electric spark, which in ▓
two points of the metal in question, vo▓
portion, and heats it so intensely as to ▓
off its peculiar light.

Thus all the metals—iron, platinum, s▓
—may each be recognized by the pecul
which its spectrum exhibits.

The permanent gases also yield charac
as hydrogen, nitrogen, oxygen, chlorine, ▓
etc.

By placing gases, solutions of salts, etc.
made for the purpose, and placing them ▓
minating flame and the slit in the tube, s
will pass through the gas or liquid, the s▓
from absorption are obtained. By this arr
portions of the complete spectrum disapp
sorption of corresponding rays of light, o▓
lines are seen in different parts of the sam

For a description of the great number o
have already been carefully studied and ma
must be made to the large works on this s

CHAPTER V.

SYSTEMATIC METHODS FOR THE DETERMINATION OF INORGANIC COMPOUNDS.

THE careful observer, having become well acquainted with the reactions which are exhibited by the metallic oxides and other simple compounds when subjected to the various treatments detailed in the second chapter, will find no difficulty in ascertaining the nature of any mineral substance presented to him for analysis.

If the reactions are not quite distinct, owing to an intermixture with other substances, he may call to his aid the processes laid down in the third chapter, which will enable him in most cases to detect also the nature of the impurities. But in order to obtain satisfactory results in this way, a certain familiarity with all the principal tests is a necessary condition; this once acquired, any further directions are quite superfluous.

Those, however, who have not devoted much time to blowpipe operations, will sometimes experience some difficulty in drawing the correct conclusions from the observed phenomena—a difficulty which is to a great extent obviated by pursuing some methodical course as given in the following schemes.

The first is applicable for all substances, but the second will be found shorter after a little experience in the case of metallic compounds.

SYSTEMATIC METHOD OF EXAMINATION OF COMPOUND SUBSTANCES.

By J. LANDAUER.

ZEITSCHRIFT FÜR ANALYTISCHE CHEMIE, XVI. 385.

PRELIMINARY EXAMINATION.

A. The substance heated in a matrass or a tube closed at one end.

a. **Gaseous substances given off:**

1. *Colorless and odorless.*

   Water: Water of crystallization; hydrate.

   Oxygen: Peroxides, nitrates, chlorates, bromates, and iodates.

   Carbon dioxide: Carbonates and oxalates.

   Carbon monoxide: Oxalates and formates. (The latter carbonize.)

2. *Colorless with odor.*

   Sulphur dioxide: Sulphites and some sulphates.

   Sulphuretted hydrogen: Thiosulphates and some hydrated sulphides.

   Ammonia: Some ammonium salts.

3. *With color and odor.*

   Nitrogen tetroxide: Most of the nitrates and nitrites.

   Iodine (violet): Some iodides and iodates.

   Bromine (brown): Some bromides.

   Chlorine (greenish-yellow): Some chlorides.

b. **Sublimate formed:**

1. *White sublimate.*

   Ammonium salts.

   Mercurous chloride, sublimes without first fusing.

   Mercuric chloride, first fuses.

Antimony oxide, fuses and forms brilliant needles.

Tellurium dioxide, fuses and sublimes to an amorphous mass.

Arsenic trioxide, octahedral crystals without fusion.

2. *Black or gray sublimate.*

Arsenic: Metallic arsenic and many of its compounds (metallic mirror).

Mercury, amalgam, and some compounds of mercury (metallic globules).

3. *Colored sublimates.*

Sulphur, yellowish-brown while hot; yellow when cold.

Antimony sulphide, black while hot; reddish-yellow when cold.

Arsenic sulphide, brownish-red while hot; reddish-yellow when cold.

Mercuric iodide, yellow; when rubbed, becomes red.

Mercuric sulphide, black; by friction, red.

Selenium, reddish to black; powder dark-red.

Change of color:

Zinc oxide, from white to yellow; when cold, white.

Tin oxide, yellowish-brown; cold, light-yellow.

Lead oxide, brownish-red; cold, yellow.

Bismuth oxide, white to orange; cold, lemon-yellow.

Mercuric oxide, red to black; cold, red (volatile).

Ferric oxide, red to black; cold, red (*not* volatile).

Mercuric iodide, red to yellow; cold, red.

Hydrated salts of cobalt, nickel, iron, and copper.

Fuse: Alkaline salts.

Carbonize: Organic substances.

f. Phosphoresce: Alkaline earthy
oxides.

g. Decrepitate: Alkaline chlorates, g
minerals.

B. The substance heated in
(When the reaction is the same as in preceding

a. Gaseous substances given off:
Sulphur dioxide, with characteris
and sulphides.
Selenium dioxide, odor of rotten l
nium and selenides.

b. Sublimate formed:
Arsenic trioxide, very volatile, wl
mote from assay; arsenic and
Antimony oxide, white fumes, ■
volatile; antimony and its con
Tellurium dioxide, white fumes, ■
colorless drops; tellurium and

Lead sulphate,   { white, mostl
Bismuth sulphate, { say; comp
       { with lead

C. The substance heated on
a. Fusibility:
1. *Fusible.* Alkalies and some salts
Antimony, lead, cadmium, te
zinc, tin (all easily fusible).
Copper, silver, gold (difficultly fu
2. *Infusible.* Salts of the earths, of t
and also silica, iron, cobalt, nic
tungsten, platinum, palladium,
and osmium.

b. **Detonation.** Nitrates, chlorates, iodates, bromates.

c. **Intumescence,** giving off water; borates and alum.

d. **Flame-coloration, reduction to metal,** and **formation of coating** are described in special reactions.

COMPLETE EXAMINATION.

DETECTION OF BASES.

The substance is treated on charcoal in the reducing flame with sodium carbonate; or in the case of a metal or alloy, alone.

a. The substance forms a coating, . See Sect. I.,   1–9.

b. The substance forms a metallic bead, without coating, .   .   " " I., 10–12.

c. The substance leaves a gray or black residue,   .   .   . " " II., 13–31.

d. The substance colors the flame, especially when moistened with HCl, ... " " IV., 32–42.

e. The substance leaves a white luminous residue, .   .   . " " V., 43–51.

f. The substance completely volatilizes, ... " " VI., 52, 53.

(Formation of hepar is indication of a sulphate or a sulphide.)

Section I.

1. *Coating white,* very volatile; disappears with light-blue flame, and characteristic alliaceous odor,

Arsenic.

1* Special Test. Heated with potassium cyanide and soda in a matrass, forms a metallic mirror.

2. *Coating reddish-brown,* with variegated border; volatilized by both flames without coloring them,

Cadmium.

15*

2* Sp. T.   The coating, placed in a matrass with sodium thiosulphate, becomes yellow.   Compare with No. 3* in presence of zinc.

3. *Coating yellow* while **hot**; *white* **when cold**; luminous and is not volatilized, ... **Zinc.**

3* Sp. T.   The coating, moistened with cobalt solution and heated, becomes green.   In presence of both Cd and Zn the Cd coating appears first, and then the Zn.

4. *Coating steel-gray;* disappears in the reducing flame with azure coloration, and gives off the odor of rotten horse-radish, ... **Selenium.**

4* Sp. T.   Compare No. 5*.

5. *Coating white,* with dark-yellow to reddish border; disappears in the reducing flame with green coloration, ... **Tellurium.**

5* Sp. T.   If Se and Te are both present, a white coating is formed, the reducing flame is colored bluish-green, and the rotten horse-radish odor given off.   For separation, a metallic coating of the two metals is made in a small glass tube, moistened with a drop of H_2SO_4, and *gently* heated.   Te dissolves with a carmine-red color, and if *strongly* heated, the dirty-green color of the Se appears.

6. *Coating bluish-white;* volatile; driven about by the oxidizing flame; disappears in the reducing flame with green coloration.

Bead: white, brittle, oxidizable,   . **Antimony.**

6* Sp. T.   If the coating is placed on platinum foil with a piece of zinc and moistened with HCl, a black adherent coating of antimony is formed on the foil.

7. *Coating orange* while hot; *lemon-yellow* when cold, driven about by both flames without coloring them.

Bead: reddish-white, brittle, oxidizable,   **Bismuth.**

7* Sp. T.   On charcoal with potassium iodide, and sulphur in the oxidizing flame, give a red coating of bismuth iodide.

8. *Coating lemon-yellow* while hot; *sulphur-yellow* when cold; driven away by the oxidizing and also reducing flame; colors the reducing flame blue.

Bead: gray, malleable, oxidizable, . . **Lead.**

8* Sp. T. Moisten the assay with HNO_3, evaporate the excess of acid, add H_2SO_4, and heat till a white vapor is given off. A white powder remains, wholly insoluble in dilute H_2SO_4.

9. *Coating yellowish* while hot; *white* when cold; very slight, close to the assay, and not volatile.

Bead: white, malleable, and very oxidizable, **Tin.**

9* Sp. Dissolves in HCl, add Zn, which precipitates metallic tin as a gray spongy mass that does not adhere to the platinum like antimony. A crystal of $Na_2S_2O_3$ precipitates brown SnS from the solution to which the Zn has been added.

10. *Bead white*, malleable, brilliant. In strong oxidizing flame, a reddish-brown coating is formed, which in the presence of lead and antimony becomes carmine-red, ... **Silver.**

10* Sp. T. Dissolved in HNO_3, HCl produces a white, curdy precipitate of AgCl.

11. *Bead yellow*, brilliant, malleable, not oxidizable,

Gold.

11* Sp. T. Dissolved in aqua regia, $SnCl_2$ gives purple of Cassius.

12. *Bead red*, malleable, and oxidizable, . **Copper.**

12* Sp. T. Compare Nos. 13 and 39.

Iron, nickel, cobalt, molybdenum, tungsten, and the platinum group of metals yield a gray infusible powder. Iron, nickel, and cobalt, which are more or less magnetic, may be further tested with borax (Sect. II.), but for the other metals the blowpipe reactions are not as characteristic. Compounds of chromium give a yellow, and of manganese, a green mass, with sodium carbonate (Section II.).

Some chlorides, iodides, bromides, and sulphides give a white coating, without a complete reduction of the metal, which should not be mistaken for any of the foregoing. These will be recognized in the course of analysis by other and more characteristic tests. Sulphuretted hydrogen, recognized by its odor, indicates a sulphate or sulphide.

Section II. Treated with borax on platinum wire.

a. Colored bead formed in the oxidizing flame and reducing flame, . . Nos. 13–31.

b. Bead not colored, . . Section IV. " 32–38.

THE COLOR OF THE BEAD IS—

| In the Oxidizing Flame. | | In the Reducing Flame. | | |
Hot.	Cold.	Hot.	Cold.	
13. Green,	Bluish-green,	Colorless,	Brown,	
14. Blue,	Blue,	Blue,	Blue,	Copper.
15. Violet to black,	Reddish-violet,	Colorless,	Colorless to pink,	Cobalt. Manganese.

13* Sp. T. With salt of phosphorus and reduced with Sn, it becomes red; if it turns black it should be heated on charcoal, and the Sb and Bi separated with boric acid in the oxidizing flame.

14* Sp. T. The metal, reduced on charcoal and placed upon paper, gives, with HNO_3, a red solution, which, with HCl added, and drying, gives a green spot, which disappears when moistened with water.

15* Sp. T. Fused with sodium carbonate and nitrate, forms a green mass.

with acetic acid and potassium ferrocyanide, which produces a brown color.

19* Sp. T.   Digested with H_2SO_4 on the platinum spoon, MoO_3 turns deep-blue if a little alcohol is added, or if it is breathed upon.

20* Sp. T.   Fused with soda and nitre on platinum, forms a yellow mass.

21* Blowpipe reactions are not decisive tests.

22* Sp. T.   Fused with soda and nitre, dissolved in water, acetic acid added, $AgNO_3$ gives a yellow precipitate.

23* Sp. T.   The salt of phosphorus bead in the oxidizing flame is colorless both when hot and cold: in the reducing flame, dirty-green while hot, blue when cold, and blood-red on the addition of iron.   Compare No. 27.

24* Sp. T.   The salt of phosphorus bead is colorless in the oxidizing flame, both when hot and cold; in the reducing flame, yellow while hot, violet when cold, and blood-red on the addition of iron.   Compare No. 30.

25. Color of the bead when more than one of the oxides is present.

In the Oxidizing Flame.		In the Reducing Flame.		
Hot.	Cold.	Hot.	Cold.	
Violet to blood-red,	Brownish-violet,	Yellow,	Bottle-green,	Fe and Mn.
Plum color,	Plum color,	Bluish-green,	Blue,	Fe, Mn, Co.
Green,	Grayish-blue,	Bluish-green,	Green,	Fe, Mn, Co, Ni.
Yellowish-green,	Green,	Greenish-blue,	Blue,	Fe, Co, little Ni.
Violet-brown,	Brown,	Blue,	Blue,	Co and much Ni.
Green,	Green, blue, or yellow, according to saturation,	. . . . .		Fe and Co. Fe and Cu. Fe and Ni.

25* Sp. T. A number of beads are made by fusing the substances with borax on platinum wire, and then reducing on charcoal with the addition of a globule of lead. After blowing for some time, the bead (A) is separated from the lead globule (B) and examined.

A. The *Bead* or fragments are fused with borax on platinum wire:

a. The bead is blue, ... **Cobalt.**

b. The bead is green while hot; blue when cold in the oxidizing flame, ... **Iron and cobalt.**

c. The bead is violet to blood-red while hot, and brownish-violet when cold in the oxidizing flame; yellow while hot, bottle-green when cold in reducing flame; on charcoal reduced with tin, vitriol-green. With insufficient oxidizing flame the bead is yellow while hot, colorless when cold, . . **Manganese and iron.**

d. The bead is plum color in the oxidizing flame, both when hot and cold; in the reducing flame, bluish-green while hot, and blue when cold,

Manganese, iron, and cobalt.

B. *The lead globule* is treated with boric acid on charcoal in the oxidizing flame to separate the lead, and the residue fused with salt of phosphorus:

a. The bead is blue when cold in the oxidizing flame, but reduced with Sn on charcoal, becomes red, **Copper.**

b. The bead is yellow when cold in oxidizing flame,

Nickel.

c. The bead is green when cold in oxidizing flame,

Copper and nickel.

Section III. The substance is fused with acid potassium sulphate, dissolved in hydrochloric acid, and a strip of zinc added.

(This section is passed by when tungsten, vanadium, titanium, and niobium are absent.)

The solution becomes—

26. Blue, then green, finally dark-brown, **Molybdic acid.**

26* Sp. T.   See No. 19.   Gree

27. Blue, then copper-red,   .   . Tungstic acid

27* Sp. T.   See No. 23.

28. Blue, then green, finally violet,   . Vanadic acid   V

28* Sp. T.   See No. 22.

29. Green, ... Chromic acid

29* Sp. T.   See No. 20.

30. Violet, ... Titanic acid

30* Sp. T.   See No. 24.

31. Blue, or in strongly acid solutions, brown,

Niobic acid.

Section IV. The substance is placed in the non-luminous flame, with the platinum forceps or wire.

a. The flame is colored, especially if the substance is moistened with HCl or H_2SO_4,   . Nos. 32–42.

b. The flame is not colored,   Section V., Nos. 43–51.

EXAMINATION FOR BASES.

The color of the flame appears—

		Alone.	Through Blue Glass.	Through Green Glass.		
Moistened with H_2SO_4, and held for a short time only in the flame.	32.	Violet.	Reddish-violet.	Bluish-green.	Potassium.	Ba, Ca, and Sr can be recognized when together, if moistened with HCl and the flashes of color in the flame carefully noticed.
	33.	Orange.	Reddish-violet.	Orange-yellow.	Potassium and sodium.	
	34.	Orange.	Invisible or pale-blue.	Orange-yellow.	Sodium.	
	35.	Carmine.	Violet-red.	Invisible.	Lithium.	
Repeatedly moistened with H_2SO_4, dried, and intensely heated.	36.	Yellowish-green.	Bluish-green	Green.	Barium.	
	37.	Yellowish-red.	Greenish-gray.	Celery-green.	Calcium.	
	38.	Carmine.	Purple.	Pale-yellow.	Strontium.	

39. Green, moistened with HCl, blue, . . **Copper.**

EXAMINATION FOR ACIDS.

40. Yellowish-green, similar to barium flame,

Molybdic acid.

40* Sp. T. Gives with borax the reactions of No. 19.

41. Yellowish-green (salts moistened with H_2SO_4),

Phosphoric acid.

41* Sp. T. Heated with a piece of Mg wire, or of Na, in the closed tube, the mass, moistened with H_2O, gives the smell of phosphoretted hydrogen.

42. Green (salts moistened with H_2SO_4), **Boric acid.**

42* Sp. T. With $CaFl_2$ and $HKSO_4$, heated on platinum, gives the intense green flame of boron fluoride.

REMARKS.—Chlorides and nitrates give green flashes, but very weak, and quickly disappear.

The flame-colors of As, Sb, Pb (blue), Zn (greenish-white), are for the most part obscured by the use of concentrated H_2SO_4.

Section V. The substance is placed on charcoal, moistened with cobalt solution, and strongly heated.

43. Blue, infusible mass, ... **Alumina.**

43 Sp. T. With No. 41, no coloration; also no silica skeleton is formed in the salt of phosphorus bead.

44. Blue, infusible mass, . . **Earthy phosphates.**

44* Sp. T. In No. 41, yellowish-green flames.

45. Blue, infusible mass, . . **Earthy silicates.**

45* Sp. T. In salt of phosphorus bead, forms a silica skeleton.

46. Blue glass, ... **Alkaline borates.**

46 Sp. T. With No. 42 shows a bright-green flame.

47. Blue glass, . . . **Alkaline phosphates.**

16

47 Sp. T. With No. 41 shows a yellowish-green flame.

48. Blue glass, ... **Alkaline silicates.**

48* Sp. T. In the salt of phosphorus bead forms a skeleton of silica.

49. Flesh-red mass, ... **Magnesia.**

50. Violet mass, ... **Zirconia.**

51. Green mass,   .   . Oxides of $\begin{cases} \textbf{Zinc,} \\ \textbf{Tin,} \\ \textbf{Antimony,} \\ \textbf{Titanium.} \end{cases}$

Section VI. The substance heated in a closed tube with soda.

52. Metallic sublimate; may be collected into small globules, ... **Mercury.**

52* Sp. T. Heated in the closed tube with $Na_2S_2O_3$ forms black HgS.

53. Smell of ammonia,   .   .   . **Ammonia.**

53* Sp. T. With HCl forms a white cloud.

DETECTION OF ACIDS.

substance heated in a closed
potassium sulphate.

gas, ...	Nos. 54–59.
pungent gas, .	" 60–67.
and inodorous gas, "	68–70.
Section VIII., "	71.

smell of nitrogen tetroxide,

Nitrous or nitric acid.

moistened with solution of ferrous
with sulphuric acid, placed in the
nitrates, heated on platinum
cyanides, detonate and burn.

Yellowish-green gas, with smell of chlorine,
Chloric acid.

55* Sp. T. Detonates on charcoal.

Violet vapor; colors starch-paste blue, . **Iodine.**

56* Sp. T. A salt of phosphorus bead, with copper oxide and an iodide compound, gives a clear-green flame.

Same as 56; with the addition of ferrous sulphate, indicates . . . . . **Iodic acid.**

57* Sp. T. The substance detonates on charcoal.

Reddish-brown vapor; colors starch-paste yellow,
Bromine.

58* Sp. T. A salt of phosphorus bead, with copper oxide and compound of bromine, colors the flame greenish-blue.

The same reactions, . . . **Bromic acid.**

59* Sp. T. The substance detonates on charcoal.

Vapors, which form with NH_3 a white cloud, and have the odor of . . **Hydrochloric acid.**

60* Sp. T. A salt of phosphorus bead, with copper oxide and compound of chlorine, colors the flame intense blue.

Strongly-fuming gas, which etches glass,
Hydrofluoric acid.

Smell of **sulphuretted hydrogen,**
Hydrosulphuric acid.

62* Sp. T. Metallic sulphides, heated in the open tube, give off sulphur dioxide fumes, which may be known by the smell and action upon moist blue litmus-paper.

Smell of burning sulphur; no separation of sulphur,
Sulphur dioxide.

Same reaction, with separation of sulphur,
Thiosulphuric acid.

65. Pungent gas; irritates the eyes to tears, and renders lime-water turbid,   .   .   . **Cyanic acid.**

66. Smell of vinegar, ... **Acetic acid.**

67. Smell of prussic acid,   .   . **Hydrocyanic acid.**

68. The gas effervesces and causes turbidity in lime-water, ... **Carbon dioxide.**

69. The gas burns with a blue flame,

Carbon monoxide.

70. Carbonization,   .   .   .   **Organic acids.**

REMARK.—Some few organic acids do not carbonize—*i. e.*, oxalic, formic, etc.

Section VIII. A substance which indicates a sulphide with soda on coal is heated in a platinum spoon with caustic potash, and the whole placed in a vessel with water, and a bright silver coin laid upon it.

71. The coin is not blackened,   . **Sulphuric acid.**

71ª Sp. T. In order to distinguish sulphates from sulphides (No. 62), the substance is dissolved in water acidified with nitric acid, and the sulphuric acid is thrown down with barium chloride.

Insoluble sulphates are first boiled in solution of sodium carbonate, then filtered, and decomposed with nitric acid.

Section IX. Already found in the course of analysis:

72. **Phosphoric acid** (No. 41), **boric acid** (No. 42), **silicic acid** (No. 45).

II. SYSTEMATIC METHOD OF EXAMINATION OF COMPOUND SUBSTANCES.

By Prof. T. EGLESTON.

American Chemist, 1872.

The substance may contain As, Sb, S, Se, Fe, Mn, Cu, Co, Ni, Pb, Bi, Ag, Au, Hg, Zn, Cd, Sn, Cl, Br, I, CO_2, SiO_2, HNO_3, H_2O, etc.

Treat on charcoal in the oxidizing flame to find volatile substances, such as As, Sb, S, Se, Pb, Bi, Cd, etc. (par. 25, *et seq.*).

Volatile substances not present.

Divide a part of the substance into three portions and proceed as in **A**.

Volatile substances present.

(1) Form a coating on charcoal and test with salt of phosphorus and tin for antimony (par. 65), or separate lead and bismuth, as in pars. 64, 74.

a. Yellow coating, yielding, with salt of phosphorus, a black bead; disappearing with blue flame, no part of it yielding green Sb flame; **Pb, Bi.**

b. Yellow coating, generally with white border, yielding black or gray bead with salt of phosphorus; disappearing with blue flame; also the border disappearing with green flame; **Pb, Sb.**

c. Yellow coating, very similar to *b*, but yielding no blue flame; **Bi, Sb.**

(2) If **As, Sb, S, Se** are present, roast a large quantity thoroughly on charcoal until no odor of arsenic or sulphur dioxide is given off. Divide the substance into three portions, and proceed as in **A**.

A. Treatment of the first portion.

Dissolve a small quantity in borax on platinum wire in the oxidizing flame, and note the color. When several oxides are present, successive colors often appear; in this case, saturate the bead and toss it off into a porcelain dish (par. 3 and page 54). Prepare several beads in this way, and treat them on charcoal with metallic lead, silver, or gold in a strong reducing flame (pars. 87, 89). If the mass spreads over the charcoal, continue blowing until a bead is formed.

The *metallic bead* (a) is removed from the *borax residue* (b) whilst hot, or with a hammer when cold, all fragments being carefully preserved.

a. The *metallic bead* contains the reduced Ni, Cu, Ag, Au, Sn, Pb, and Bi (Sn, Pb, and Bi are partially volatilized).

Treat this bead on charcoal in the oxidizing flame until all the Pb is removed, or remove the Pb with boric acid (par. 88). Ni, Co, Ag, Au remain behind.

Treat this residue on charcoal (oxidizing flame) with salt of phosphorus, and remove the bead whilst hot:

A green bead when cold indicates (par. 89) Ni and Cu, yellow Ni, blue Cu.

The Cu bead, heated on charcoal (reducing flame) with metallic tin, becomes red (par. 88).

The presence of Ag and Au is ascertained by special examination.

b. The *borax residue* retains the Fe, Mn, Co, etc.

Dissolve a fragment of the bead in borax on the platinum wire; a blue bead indicates,   Co.

In presence of much iron add more borax to detect the Co (pars. 86, 87).

Dark-violet or black bead in oxidizing flame,   Mn.

When only Fe and Mn are present, an almost colorless bead (reducing flame) results.

Test in the wet way (par. 51) for **Cr, Ti, Mo, Nb, W, V.**

B. Treatment of the second portion.

Heat on charcoal in the reducing flame with Na_2CO_3, and look for indications of **Zn, Cd,** and **Sn.** If a white coating results, treat with cobalt solution (par. 60).

C. Treatment of the third portion.

Dissolve in salt of phosphorus on platinum wire (oxidizing flame) for **SiO_2**, and test for **Mn** with KNO_3 (par. 104).

Special tests.

1. To confirm **As,** heat on charcoal with Na_2CO_3, or in closed tube with *dry* Na_2CO_3 (par. 69, *et seq.*).

2. Dissolve in salt of phosphorus on platinum wire in the oxidizing flame (provided the assay is neither a metal nor contains S), and test for **Sb** on charcoal with metallic tin in reducing flame (par. 65).

3. Test for **Se** on charcoal (par. 113).

4. In absence of Se, fuse with Na_2CO_3 (reducing flame), and test for **S** on silver foil (par. 121). If **Se** be present, test for **S** in an open tube (par. 20). To distinguish between sulphides and sulphates see par. 104.

5. Test for **Hg** by heating in closed tube with *dry* Na_2CO_3 (par. 11).

6. Fuse with assay-lead and borax-glass on charcoal in reducing flame. Cupel the Pb bead for **Ag** (par. 117). Test for **Au** by means of HNO_3 (par. 95).

7. Test for **Cl, Br,** and I with salt of phosphorus bead containing Cu (pars. 82, 97).

8. Test Cl and Br with $HKSO_4$ (par. 79).
9. Test for water in closed tube (par. 9).
10. Apply flame-coloration tests (par. 57–61).
11. Test for CO_2 with HCl.
12. Test for HNO_3 by means of $HKSO_4$ (par. 106).
13. Test for Te according to par. 122.

When sulphides, etc. are under examination, they must be well roasted ; but if S, As, Sb, or Se, as sulphides, etc., are absent, the substance is either an oxide or an alloy. If an oxide, the roasting is unnecessary; if an alloy, it is tested by (1) *a* for Pb, etc., and then the test is made by fusing the substance on charcoal with borax in the reducing flame, thus performing in the same operation the test A and A, *a*. Some sulphides during roasting (A) are reduced, and are then treated like alloys. Metals, sulphides, etc. should be fused on charcoal with the flux, and not on platinum. The reducing flame is used if only the nonreducible metals, as Fe, Co, etc., are to be obtained in the flux ; the oxidizing, if Cu, Ni, etc., and other reducible metals. The flux may then be taken up on the wire. Sulphides, etc. must always be roasted before testing with borax and salt of phosphorus.

In B, Sn can always be found in the presence of Zn by reducing the oxides with soda and a little borax, and triturating the mass in water (par. 46). In some alloys, for example, bronzes, containing both Sn and Zn, the latter can be detected by treating a short time in the reducing flame, and testing the coating formed with cobalt solution, as the Zn coating forms first.

The word *button* refers to the metal, and *bead* to the flux.

TABLES

SHOWING THE BEHAVIOR OF THE ALKALIES,
EARTHS, AND METALLIC OXIDES, ALONE,
AND WITH REAGENTS, BEFORE
THE BLOW-PIPE.

TABLE I.—Behavior of the alkalies and alkaline earths before the blowpipe.

Alkalies.	*Alone on Platinum Wire.*
1. POTASH. K_2O.	Colors the flame violet; but even a minute quantity of soda obscures the reaction. See par. 58.
2. SODA. Na_2O.	Colors the flame intense reddish-yellow, even in the presence of a large excess of potash.
3. LITHIA. Li_2O.	Colors the flame carmine-red, even in the presence of potash; but soda gives a yellowish-red. See par. 59.
4. AMMONIA. NH_3.	Combined with chlorine, nitric or sulphuric acids, it colors the flame very pale-green.

TABLE I. — Continued.

Alone on Platinum Foil.	Remarks.
No change.	In solution, change red litmus paper to blue.
No change.	Same as Potash.
Turns the foil yellow when fused; but if washed and ignited the color is destroyed, but the foil remains dull.	Same as Potash.
No reaction.	Pungent odor, colors red litmus paper blue.

TABLE I.—Continued.

Alkaline Earths.	On Charcoal alone, and in the forceps.	With Sodium Carbonate on Charcoal.
5. BARYTA. BaO.	The Hydrate fuses, boils, intumesces, and is finally absorbed by the charcoal. The Carbonate fuses readily to a transparent glass, which, on cooling, becomes enamel-white. In the forceps it colors the outer flame yellowish-green.	Fuses to a homogeneous mass, which is absorbed by the charcoal.
6. STRONTIA. SrO.	The Hydrate behaves like Barium hydrate. The Carbonate fuses only on the edges, and swells out in arborescent ramifications, which emit a brilliant light, and, when heated with the R. F., impart to it a reddish tinge; shows after cooling alkaline reaction. In the forceps, colors the outer flame crimson.	Caustic Strontia is insoluble. The Carbonate, mixed with its own volume of soda, fuses into a limpid glass, which becomes enamel-white on cooling. At a greater heat the mass boils, and caustic Strontia is formed, which is absorbed by the charcoal.
7. LIME. CaO.	Caustic Lime is not changed. The Carbonate loses carbon dioxide, becomes whiter and more luminous, and shows after cooling alkaline reaction. In the forceps it colors the outer flame pale-red.	Insoluble. The soda passes into the charcoal, and leaves the lime unaltered on its surface.
8. MAGNESIA. MgO.	Undergoes no alteration. The Carbonate becomes caustic and luminous. On addition of cobalt solution and flaming becomes pink.	It behaves like lime.
9. ALUMINA. Al$_2$O$_3$.	Not changed. On addition of cobalt solution and flaming becomes blue.	Forms an infusible compound, with slight intumescence. The excess of soda is absorbed by the charcoal.

TABLE I.—Continued.

With Borax on Platinum Wire.	With Salt of Phosphorus on Platinum Wire.
The Carbonate dissolves with effervescence to a limpid glass which, with a certain amount, becomes opaque by flaming; with more, it becomes opaque-white on cooling, even without flaming.	As with Borax.
Same as Baryta.	Same as Baryta.
Readily dissolved to a limpid glass, which becomes opaque by flaming. The Carbonate dissolves with effervescence. On a large addition of Lime the glass becomes cloudy and crystallizes on cooling, but does not become enamel-white, like Baryta or Strontia.	Soluble in large quantities to a limpid glass which, when sufficient Lime is present, becomes opaque by flaming. When saturated, the glass becomes enamel-white on cooling.
It behaves like lime, but not so crystalline.	Readily soluble to a limpid glass, which becomes opaque by flaming. When saturated, it becomes, on cooling, enamel-white.
Dissolves slowly to a limpid glass, which remains so on cooling, and which cannot be made cloudy by flaming. A large quantity of alumina makes the glass cloudy and nearly infusible; on cooling, it then assumes a crystalline surface, and is scarcely fusible.	Soluble to a limpid glass, which remains clear under all circumstances. If too much alumina be added, the undissolved portion becomes translucent.

TABLE I.—Continued.

Alkaline Earths.	On Charcoal alone, and in the forceps.	With Sodium Carbo on Charcoal.
10. GLUCINA. BeO.	Not changed. With cobalt solution and flaming turns bluish-gray.	Insoluble.
11. YTTRIA. YO.	Not changed.	Insoluble.
12. ZIRCONIA. Zr^2O^3.	Infusible, but emitting a very glaring light.	Insoluble.
13. ERBIA EO.	The yellow oxide becomes lighter-colored and transparent in the R. F.	Insoluble.
14. THORIA. ThO.	Not changed.	Insoluble.
15. SILICA. SiO_2.	Not changed. On addition of cobalt solution and flaming becomes pale-blue. If slightly fused the color is deeper.	Soluble, with effi cence, to a clear gl

TABLE I.— Continued.

With Borax on Platinum Wire.	*With Salt of Phosphorus on Platinum Wire.*
Soluble in large quantities to a limpid glass, which becomes opaque by flaming. When Glucina is present in excess, it becomes enamel-white on cooling.	As with Borax.
Like Glucina.	Like Glucina.
Like Glucina.	Dissolves more slowly than with Borax.
Dissolves slowly to a clear glass, which becomes opaque by flaming, or on cooling if in excess.	As with Borax.
In small quantity dissolves to a clear glass, which becomes enamel-white on cooling if in excess; if clear, it cannot be made opaque by flaming.	As with Borax.
Dissolves slowly to a limpid glass, difficultly fusible, and cannot be made opaque by flaming.	Soluble in very small quantities to a limpid glass. The insoluble portion, or silica skeleton, floats about as a translucent mass in the clear glass.

TABLE II.—Behavior of the metallic oxides before the blowpipe.

Metallic Oxides in Alphabetical Order.	On Charcoal alone.	With Sodium Carbonate on Charcoal.
1. ANTIMONY TRIOXIDE. Sb_2O_3.	O. F.: It is displaced and deposited upon another part of the charcoal. R. F.: It is reduced and volatilized. A coating of oxide is deposited on the charcoal, and a greenish-blue color imparted to the flame.	On charcoal very readily reduced in O. F. and R. F. The metal fumes and coats the charcoal with antimony oxide.
2. ARSENIC TRIOXIDE. As_2O_3.	Volatilizes below a red heat.	On charcoal reduced, with emission of arsenical fumes, which are characterized by a strong garlic odor.
3. BISMUTH TRIOXIDE. Bi_2O_3.	O. F.: On platinum foil it fuses readily to a dark-brown mass, which, on cooling, becomes pale-yellow. On charcoal in O. F. and R. F. reduced to metallic bismuth, which, with long blowing, vaporizes, coating the charcoal with yellow oxide. The coating, when touched with the R. F., disappears without coloring the flame.	Easily reduced to metallic bismuth.
4. CADMIUM OXIDE. CdO.	O. F.: On platinum foil unchanged. R. F.: On charcoal it disappears in a short time, and deposits all over the charcoal a dark-yellow or reddish-brown powder. The outer part of the coating is also iridescent.	O. F.: Insoluble. R. F.: On charcoal readily reduced; the metal vaporizes and deposits a dark-yellow or reddish-brown coating on the charcoal. The more remote portion of the coal assumes a variegated appearance.

TABLE II.—Continued.

With Borax on Platinum Wire.	*With Salt of Phosphorus on Platinum Wire.*
O. F. : Dissolves in large quantities to a limpid glass, which, while hot, appears yellowish, but after cooling, colorless. R. F. : The glass, when treated only for a short time in the O. F., becomes on Ch. grayish and cloudy from particles of reduced antimony. With tin it becomes gray or black, according to the degree of saturation.	O. F. : Dissolves with effervescence to a limpid glass, which, while hot, is slightly yellowish. R. F. : On Ch. the saturated bead becomes at first cloudy, but afterwards clear again, owing to the volatilization of the reduced antimony. Treated with tin, the glass becomes, after cooling, gray, even if but very little antimony trioxide is present. With strong blowing it becomes clear again.
O. F. : A small quantity is easily dissolved to a clear yellow glass, which, on cooling, becomes colorless. On a large addition of oxide, the glass, while hot, is yellowish-red, becomes yellow on cooling, and when cold is opalescent. R. F. : On Ch. the glass becomes at first gray and cloudy, the oxide is reduced to metal with effervescence, and the bead becomes clear again. An addition of tin accelerates the process.	O. F. : Readily dissolved to a limpid yellow glass, which, on cooling, becomes colorless. When a greater quantity of oxide is present, the glass may be made enamel-white by flaming, and on a still larger addition it becomes by itself enamel-white on cooling. R. F. : On Ch., particularly when tin is added, the glass remains colorless and limpid while hot, but becomes, on cooling, dark-gray and opaque.
O. F. : Soluble in large quantity to a limpid yellowish glass, becoming almost colorless on cooling. When highly saturated, it may be made enamel-white by flaming, and when still more oxide is present, it becomes by itself enamel-white on cooling. R. F. : Placed on Ch., it enters into ebullition; the oxide is reduced; the reduced metal vaporizes immediately and deposits a dark-yellow coating.	O. F. : Soluble in large quantity to a limpid glass, which, while hot, is yellowish, but colorless when cold; when saturated, it becomes enamel-white on cooling. R. F. : On Ch., the oxide becomes slowly and imperfectly reduced. The reduced metal deposits a very feeble Ct. of dark-yellow color. The color is only clearly seen when the mass is cold. An addition of tin facilitates the reduction.

TABLE II. — Continued.

Metallic Oxides in Alphabetical Order.	On Charcoal alone.	With Sodium Carbonate on Charcoal.
5. CERIUM SESQUIOXIDE. Ce_2O_3.	The protoxide is converted into sesquioxide, Ce^2O^3 by the O. F., which remains unaltered in the R. F.	Insoluble. The soda passes into the charcoal; the sesquioxide is reduced to protoxide, which remains on the charcoal as as a light-gray powder.
6. CHROMIUM SESQUIOXIDE. Cr_2O_3.	Not changed in the O. F. or R. F.	O. F. : On platinum wire soluble to a dark yellowish-brown glass, which on cooling becomes opaque and yellow. (Chromic acid.) R. F. : The glass becomes opaque and green on cooling. On charcoal it cannot be reduced to metal; the soda passes into the charcoal, and the oxide remains behind as a green powder, Cr_2O_3.
7. COBALT MONOXIDE. CoO.	O. F.: Not changed. R. F.: It is reduced to metal, but does not fuse: the mass is attracted by the magnet, and assumes metallic lustre by friction.	O. F.: On platinum wire a very small quantity is dissolved to a transparent mass of a pale-reddish color, which, on cooling, becomes gray. R. F.: On charcoal reduced to a gray magnetic powder, which becomes lustrous by rubbing.

TABLE II. — Continued.

With Borax on Platinum Wire.	With Salt of Phosphorus on Platinum Wire.
O. F. : Soluble to a limpid glass of dark-yellow or red color, which changes on cooling to yellow. When highly saturated with oxide the glass becomes, on cooling, enamel-white. R.F. : The yellow glass becomes colorless. A highly saturated bead becomes on cooling enamel-white and crystalline.	O. F.: As with borax, but on cooling, colorless. R. F.: Perfectly colorless, hot and cold, thus being distinguished from an iron sesquioxide glass. Never becomes opaque on cooling, however large the amount of oxide present.
O. F. : Dissolves but slowly, but colors intensively. If little of the oxide is present, the glass, while hot, is yellow; when cold, yellowish-green; with more oxide it is dark-red while hot, becomes yellow on cooling, and when perfectly cold has a fine yellowish-green color. R. F. : The glass is green, hot and cold. The intensity of the color depends on the amount of oxide present. Tin causes no change.	O. F.: Soluble to a limpid glass, which, while hot, appears reddish; when cold it has a fine green color. R. F.: As in O. F., but the colors are more intense. The same on addition of tin.
O. F. : Colors very intensively. The glass appears pure smalt-blue, hot and cold. An excess of oxide imparts to the bead a deep bluish-black color. R. F. : As in O. F.	O. F.: As with borax, but for the same quantity of oxide the color is not quite so deep. R. F.: As in O. F.

Metallic Oxides in Alphabetical Order.	*On Charcoal alone*
8. COPPER OXIDE. CuO.	O. F.: Fuses to a b globule, which bec reduced where it is in tact with the charcoal R. F.: Reduced to n at a temperature belov melting-point of co When the heat is in ed, a globule of copper is obtained.
9. DIDYMIUM OXIDE. D_2O_3.	Unaltered.
10. GOLD TRIOXIDE. Au_2O_3.	When heated to tion it becomes reduc metal in O. F. and The metal fuses eas a globule.
11. INDIUM OXIDE. In_2O_3.	In the O. F. bec dark-yellow when he but on cooling is lighter, and does not In the R. F. is grad reduced and volatil depositing a coat o coal. A distinct flame is produced.
12. IRIDIUM DIOXIDE.	At a red heat bec reduced; the red

TABLE II.—Continued.

With Borax on Platinum Wire.	With Salt of Phosphorus on Platinum Wire.
O. F. : A small addition of oxide makes the glass appear green while hot, but blue when cold. A large quantity imparts to it a very deep-green color while hot, becoming greenish-blue when cold. R. F. : A glass containing a certain quantity of oxide becomes colorless, but on cooling becomes opaque and red (suboxide). On Ch. the copper may be precipitated in the metallic state, the bead becoming in consequence colorless. A glass containing protoxide, when treated on Ch. with tin, becomes on cooling brownish-red and opaque.	O. F.: As with borax, but for the same amount of oxide the coloration is not so deep. R. F. : A glass containing a large quantity of oxide becomes dark-green, which in the moment of refrigeration changes suddenly to brownish-red and opaque. A glass containing but little oxide, when treated on Ch. with tin, appears colorless while hot, but becomes brownish-red and opaque on cooling.
In the O. F. soluble to a clear colorless glass, and remains unaltered in the R. F. When strongly saturated, becomes rose-color.	Dissolves with more difficulty than in borax, but when strongly saturated is distinctly rose-red in the R. F.
As with sodium carbonate.	As with sodium carbonate.
In the O. F. dissolves to a clear glass, feebly yellowish while hot, colorless on cooling, and cloudy when much is added. In the R. F. the glass is unchanged. On coal the oxide is reduced, volatilizes, and coats the coal again with oxide. The flame is violet even in presence of soda.	As with borax, but the glass, when treated with tin on coal, becomes gray and cloudy on cooling.
As with sodium carbonate.	As with sodium carbonate.

TABLE II. — Continued.

Metallic Oxides in Alphabetical Order.	On Charcoal alone.	With Sodium Carbonate on Charcoal.
13. IRON SESQUIOXIDE. Fe_2O_3.	O. F. : Not changed. R. F. : Becomes black and magnetic, Fe_3O_4.	O. F. : Insoluble. R. F. : On Ch. it is reduced; the mass, when placed in a mortar, pulverized, and repeatedly washed with water to remove the adherent Ch. particles, yields a gray metallic powder which is attracted by the magnet, Fe_3O_4.
14. LANTHANUM SESQUIOXIDE. La_2O_3.	Unchanged.	Insoluble. The soda is absorbed by the coal, leaving the gray oxide behind.
15. LEAD OXIDE. PbO.	Minium, when heated on platinum foil, blackens; on increasing the temperature it changes into yellow oxide, which finally fuses to a yellow glass. On Ch. in O. F. and R. F. almost instantaneously reduced to metal which, with continued blowing, vaporizes, and covers the Ch. with yellow oxide, surrounded by a faint white ring of carbonate. The Ct., when touched with the R. F., disappears, imparting to the flame an azure-blue tinge.	O. F. : On platinum wire readily dissolved to a limpid glass, which, on cooling, becomes yellowish and opaque. R. F. : On Ch. reduced to metal which, with continued blowing, covers the Ch. with oxide.

TABLE II.—Continued.

With Borax on Platinum Wire.	*With Salt of Phosphorus on Platinum Wire.*
O. F. : A small amount of oxide causes the glass to look yellow while hot, colorless when cold. When more of the oxide is present the glass, while hot, appears red, and yellow when cold. A still larger quantity makes the glass dark-red while hot, and dark-yellow when cold. R. F. : The glass becomes bottle-green. Treated on Ch. with tin it becomes, at first, bottle-green, but afterwards pure vitriol-green.	O. F. : When at a certain point of saturation the glass, while hot, appears yellowish-red, and becomes on cooling at first yellow, then greenish, and finally colorless. On a very large addition of oxide it appears, while hot, deep-red, becoming, on cooling, brownish-red, then of a dirty-green color, and finally less brownish-red. R. F. : A glass containing but little of the oxide suffers no visible change. When more of the oxide is present it is red while hot, and on cooling becomes at first yellow, then greenish, and finally reddish. Treated with tin on Ch. the glass on cooling becomes at first green, and finally colorless.
In the O. F. dissolves to a clear, colorless glass, that becomes enamel-white by flaming when saturated to a certain extent, and when strongly saturated becomes enamel-like of itself on cooling. In R. F. the same as in O. F.	As with Borax.
O. F. : Easily soluble to a limpid yellow glass which, on cooling, becomes colorless. If much oxide be present it may be made cloudy by flaming. A still larger addition of oxide causes the bead to become enamel-yellow on cooling. R. F. : The glass diffuses itself over the Ch. and becomes cloudy. With continued blowing the oxide is reduced to metal, with effervescence, and the glass becomes clear again.	O. F.: As with borax. But to obtain a glass which appears yellow while hot, a large addition of the oxide is required. R. F. : On Ch. the glass becomes grayish and cloudy. This phenomenon is better observed when tin is added; but the glass can never be made quite opaque. If much of the oxide be present, the Ch. becomes coated.

TABLE II.—Continued.

Metallic Oxides in Alphabetical Order.	On Charcoal alone.	With Sodium Carbonate on Charcoal.
16. MANGANESE DIOXIDE. MnO_2.	O. F. : Infusible. When the temperature is sufficiently high, both the sesquioxide and the peroxide are converted into a reddish-brown powder, Mn_3O_4. R. F. : The same effect.	O. F. : On platinum wire or foil a very small quantity dissolves to a transparent green mass, which on cooling becomes opaque and bluish-green. R. F. : On charcoal it cannot be reduced to metal; the soda passes into the charcoal and leaves the oxide behind.
17. MERCURY OXIDE. HgO	Instantly reduced and volatilized.	Heated in a matrass to redness, it is reduced and vaporized. The vapors condense in the neck of the matrass and form a metallic coating, which can be united to a globule by carefully tapping on the matrass.
18. MOLYBDE- NUM TRIOXIDE. MoO_3.	O. F. : Fuses, becomes brown, vaporizes, and deposits on the Ch. a yellow Ct., which nearest to the assay is crystalline. On cooling the Ct. becomes white, and the crystals colorless. Beyond this coat is a thinner non-volatile film of dioxide, which on cooling is dark copper red, with metallic lustre. R. F. : The greater part of the assay is absorbed by the Ch., and may be reduced to metal at a sufficiently high temperature; the metal is in the shape of a gray powder.	O. F. : On platinum wire dissolves with effervescence to a limpid glass, which on cooling becomes milk-white. R. F. : Fuses with effervescence. The fused mass is absorbed by the Ch., and part of the acid is reduced to metal which may be obtained as a steel-gray powder.

TABLE II. — Continued.

With Borax on Platinum Wire.	With Salt of Phosphorus on Platinum Wire.
O. F. : Colors very intensively. The glass, while hot, is violet, on cooling it assumes a reddish tinge. When much manganese is added, the glass becomes quite black and opaque ; but the color can be seen when the glass, while soft, is flattened with the forceps. R. F. : The glass becomes colorless. If the color is very dark, the phenomenon is best observed on Ch. with addition of tin.	O. F. : A considerable addition of manganese must be made to produce a colored glass; it then appears, while hot, brownish- violet, and reddish-violet when cold, but never opaque If the glass contain so small a quantity of manganese that it appears colorless, an addition of nitre will produce the characteristic coloration. A glass containing oxide bubbles and yields gas at a high temperature. R. F. : Becomes very soon colorless, and remains then quiet.
O. F.: Dissolved in large quantities to a limpid glass, which, while hot, appears yellow, but colorless on cooling. A very large amount of oxide causes the glass to appear dark-yellow while hot and opaline when cold. R. F. : A highly saturated bead becomes brown, and opaque when still more oxide is present. In a good flame black flocks of MoO_2 separate, and are visible in the yellow glass if flattened.	O. F.: Easily soluble to a limpid glass. If but little of the acid be present it is yellowish-green while hot, but when cold almost colorless. On the charcoal the glass becomes very dark, and on cooling assumes a beautiful green color, from the dioxide produced by the reducing action of carbon monoxide. R. F.: The glass assumes a very dark, dirty-green color, which on cooling becomes beautiful bright-green. The same on charcoal; tin deepens the color a little.

TABLE I.—Continued.

Alkaline Earths.	On Charcoal alone, and in the forceps.	With Sodium Carbonate on Charcoal.
10. GLUCINA. BeO.	Not changed. With cobalt solution and flaming turns bluish-gray.	Insoluble.
11. YTTRIA. YO.	Not changed.	Insoluble.
12. ZIRCONIA. Zr^2O^3.	Infusible, but emitting a very glaring light.	Insoluble.
13. ERBIA EO.	The yellow oxide becomes lighter-colored and transparent in the R. F.	Insoluble.
14. THORIA. ThO.	Not changed.	Insoluble.
15. SILICA. SiO_2.	Not changed. On addition of cobalt solution and flaming becomes pale-blue. If slightly fused the color is deeper.	Soluble, with effervescence, to a clear glass.

TABLE II.—Continued.

With Borax on Platinum Wire.	With Salt of Phosphorus on Platinum Wire.
O. F. : A small quantity colors the bead violet while hot; when cold, pale reddish-brown. More oxide makes the coloration deeper. R. F. : The glass becomes gray and cloudy, or even opaque. With continued blowing the minute particles of reduced metal collect together and the glass becomes colorless. This takes place more readily on Ch., especially when tin is added. The nickel then unites with the tin to a globule.	O. F. : Soluble to a reddish glass which, on cooling, becomes yellow. A larger addition causes the glass to appear brownish-red while hot, and reddish-yellow when cold. R. F. : On platinum wire not changed. On Ch. with tin it becomes, at first, gray and opaque; with continued blowing the nickel becomes reduced, and the glass clear again and colorless.
In O. F. dissolves easily to a clear, colorless glass, becoming opaque by flaming with a certain addition, and with more becomes opaque of itself when cool. In R. F. a glass which, after treatment in the O. F., becomes opaque of itself, on cooling remaining unaltered.	In O. F. dissolves largely to a clear glass, yellow while hot, but colorless on cooling. In R. F., with a large addition, the glass becomes brown. The addition of ferrous sulphate gives a blood-red bead.

TABLE II.—Continued.

Metallic Oxides in Alphabetical Order.	On Charcoal alone.	With Sodium Carbonate on Charcoal.
21. OSMIUM DIOXIDE. OsO_2.	O. F.: Converted into osmic acid, which, without depositing a coating, volatilizes with its peculiar pungent odor. R. F.: Easily reduced to a dark-brown and infusible metallic powder.	Easily reduced to an infusible metallic powder, which may be obtained pure by washing away the coal.
22. PALLADIUM MONOXIDE. PdO.	Reduced at a red heat; but the metallic particles are infusible.	Insoluble. The soda passes into the charcoal, and leaves the palladium behind as an infusible powder.
23. PLATINUM DIOXIDE. PtO_2.	Like palladium.	Like palladium.
24. RHODIUM OXIDE. R_2O_3.	Like palladium.	Like palladium.
25. RUTHENIUM OXIDE. Ru_2O_3.	Like palladium.	Like palladium.
26. SILVER OXIDE. Ag_2O.	Easily reduced to metallic silver, which unites to one or more globules.	Instantly reduced. The soda passes into the charcoal, and the metal unites to one or more globules.
27. TANTALUM PENTOXIDE. Ta_2O_5.	In O. F. becomes slightly yellow, but is white again when cold. In R. F. the same. On addition of Cobalt becomes light-gray after long ignition, and turns slightly red, like magnesia, on cooling.	In R. F., with more than an equal volume of soda, fuses on coal to a bead with effervescence, and soon spreads out; with more soda, sinks into the coal. In R. F. the same. It cannot be reduced to metal.

TABLE II.—Continued.

With Borax on Platinum Wire.	With Salt of Phosphorus on Platinum Wire.
O. F. : A small quantity colors the bead violet while hot; when cold, pale reddish-brown. More oxide makes the coloration deeper. R. F. : The glass becomes gray and cloudy, or even opaque. With continued blowing the minute particles of reduced metal collect together and the glass becomes colorless. This takes place more readily on Ch., especially when tin is added. The nickel then unites with the tin to a globule.	O. F. : Soluble to a reddish glass which, on cooling, becomes yellow. A larger addition causes the glass to appear brownish-red while hot, and reddish-yellow when cold. R. F. : On platinum wire not changed. On Ch. with tin it becomes, at first, gray and opaque; with continued blowing the nickel becomes reduced, and the glass clear again and colorless.
In O. F. dissolves easily to a clear, colorless glass, becoming opaque by flaming with a certain addition, and with more becomes opaque of itself when cool. In R. F. a glass which, after treatment in the O. F., becomes opaque of itself, on cooling remaining unaltered.	In O. F. dissolves largely to a clear glass, yellow while hot, but colorless on cooling. In R. F., with a large addition, the glass becomes brown. The addition of ferrous sulphate gives a blood-red bead.

TABLE II.—Continued.

Metallic Oxides in Alphabetical Order.	On Charcoal alone.	With
8. COPPER OXIDE. CuO.	O. F.: Fuses to a black globule, which becomes reduced where it is in contact with the charcoal. R. F.: Reduced to metal at a temperature below the melting-point of copper. When the heat is increased, a globule of metallic copper is obtained.	O. F. soluble of gre ing it b white. R. F. ly redu when sufficie one or
9. DIDYMIUM OXIDE. D_2O_3.	Unaltered.	Inso absorb the gr hind.
10. GOLD TRIOXIDE. Au_2O_3.	When heated to ignition it becomes reduced to metal in O. F. and R. F. The metal fuses easily to a globule.	Doe soda, b in both fuses r The sc charco
11. INDIUM OXIDE. In_2O_3.	In the O. F. becomes dark-yellow when heated, but on cooling is again lighter, and does not fuse. In the R. F. is gradually reduced and volatilized, depositing a coat on the coal. A distinct violet flame is produced.	In tl In tl on co partly the co a porti almost ules in
12. IRIDIUM DIOXIDE. IrO_2.	At a red heat becomes reduced; the reduced metal is infusible.	O. F in the redno be fuse R. F

TABLE II.—Continued.

: on *Platinum Wire.*	With Salt of *Phosphorus on Platinum Wire.*
small addition of s the glass appear hot, but blue when ge quantity imparts to ep-green color while g greenish-blue when glass containing a cer- y of oxide becomes t on cooling becomes red (suboxide). On er may be precipitated ic state, the bead be- onsequence colorless. ining protoxide, when h. with tin, becomes brownish - red and	O. F.: As with borax, but for the same amount of oxide the coloration is not so deep. R. F.: A glass containing a large quantity of oxide becomes dark-green, which in the moment of refrigeration changes suddenly to brownish-red and opaque. A glass containing but little oxide, when treated on Ch. with tin, appears colorless while hot, but becomes brownish-red and opaque on cooling.
F. soluble to a clear ss and remains unal- R. F. When strongly comes rose-color.	Dissolves with more difficulty than in borax, but when strongly saturated is distinctly rose-red in the R. F.
sodium carbonate.	As with sodium carbonate.
F. dissoves to a clear yellowish while hot, cooling, and cloudy is added. . F. the glass is un- On coal the oxide is atilizes, and coats the ith oxide. The flame i in presence of soda.	As with borax, but the glass, when treated with tin on coal, becomes gray and cloudy on cooling.
sodium carbonate.	As with sodium carbonate.

TABLE II.—Continued.

Metallic Oxides in Alphabetical Order.	On Charcoal alone.	Wit
28. TELLURIUM DIOXIDE. TeO$_2$.	O. F.: Fuses, and is reduced with effervescence. The reduced metal becomes instant y vaporized and covers the charcoal with tellurium dioxide; the coating usually has a red or dark-yellow edge. R. F.: As in O. F.; the outer flame appears of a bluish-green color.	Sol wire, colorl coolit On volati coatir ide.
29. THALLIUM MONOXIDE. Tl$_2$O.	Melts and is reduced with effervescence to metallic globules, which volatilize on continued blowing, and yield a slight white coating.	On to mi of var a whi
30. TIN DIOXIDE. SnO$_2$.	O. F.: The protoxide burns, like tinder, to dioxide. The dioxide becomes very luminous, and appears, while hot, yellowish, but assumes on cooling a dirty-white color. R. F.: With a powerful and continued flame it may be reduced to metal, a trifle of dioxide being formed near the metal.	O. l it forn ferve comp R. duced

TABLE II.—Continued.

With Borax on Platinum Wire.	*With Salt of Phosphorus on Platinum Wire.*
O. F.: Soluble to a limpid and colorless glass, which, on charcoal becomes gray from reduced metal. R. F.: On charcoal becomes at first gray; afterward colorless. The charcoal becomes coated with tellurium dioxide.	As with borax.
In O. F. dissolves readily to a colorless glass. When the cooled bead is heated below redness, the surface is colored brown; if the heat be raised to redness, the bead becomes again colorless after slight blowing. In the R. F. the bead acquires a grayish turbidity, which disappears on continued blowing.	In O. F. dissolves to a clear glass, which, when slightly blown upon, becomes turbid. In R. F. on charcoal the yellow glass becomes gray and clouded.
O. F.: A very small quantity dissolves slowly to a limpid and colorless glass, which remains so on cooling, and not becoming opaque by flaming. R. F.: From a highly-saturated glass a part of the oxide may be reduced on charcoal.	O. F.: As with borax. R. F.: The glass, containing oxide, suffers no change on coal or platinum wire.

TABLE II.—Continued.

With Borax on Platinum Wire.	With Salt of Phosphorus on Platinum Wire.
O. F.: Easily soluble to a limpid glass, which, when containing a large quantity, appears yellow while hot, but becomes colorless on cooling. When containing a very large quantity it is enamel-white when cold. R. F.: When containing but little titanium dioxide, the glass becomes yellow; when more, dark-yellow to brown. A saturated glass becomes enamel-blue by flaming.	O. F.: Easily dissolved to a limpid glass, which, when containing a large quantity, appears yellow while hot, but becomes colorless on cooling. R. F.: Appears yellow while hot, but on cooling reddens, and finally assumes a violet color. If iron be present, the glass, on cooling, becomes brownish-red; with tin, or if the amount of titanium be very small, metallic zinc, on charcoal, the glass becomes violet, unless the amount of iron be very considerable.
O. F.: Like titanium dioxide. R. F.: A glass containing but little tungsten trioxide is not changed. When more, it becomes yellow, and, on cooling, yellowish-brown. On charcoal the same reaction is produced with a less saturated bead. Tin deepens the colors.	O. F.: Easily dissolved to a limpid and colorless bead, which, when highly saturated, appears yellow while hot. R. F.: With little blowing the glass appears, while hot, of a dirty green color, blue on cooling; with strong blowing it becomes, on cooling, bluish-green. On charcoal with tin, deep-green. If iron be present, the glass, on cooling, becomes brownish-red; with tin on charcoal the glass becomes blue, or, if the amount of iron be considerable, green.
O. F.: Behaves like iron sesquioxide. When highly saturated the glass may be made enamel-yellow by flaming. R. F.: Behaves like iron sesquioxide. The green bead, when at a certain point of saturation, may be made black by flaming. On charcoal with tin it becomes dark-green.	O. F.: Dissolves to a limpid yellow glass, which, on cooling, becomes yellowish-green. R. F.: The glass assumes a dirty-green color, which, on cooling, changes to a fine green. With tin on charcoal the color deepens.

TABLE II.—Continued.

Metallic Oxides in Alphabetical Order.	On Charcoal alone.	
34. VANADIUM PENTOXIDE. V_2O_5.	Fusible. Where it is in contact with the charcoal it becomes reduced and passes into the charcoal. The rest assumes the lustre and color of graphite, and consists of trioxide V_2O_3.	Unites to a fusible mass which is absorbed by the charcoal.
35. ZINC OXIDE. ZnO.	O. F.: When heated becomes yellow, and, on cooling, white again. It fuses not, but becomes very luminous. R. F.: Is slowly reduced; the reduced metal becomes rapidly reoxidized and the oxide deposited on another place of the charcoal, being yellowish while hot, and white on cooling.	O. F.: Insoluble. R. F.: On charcoal it becomes reduced. The metal vaporizes and coats the charcoal with oxide. With a powerful flame the characteristic zinc flame is sometimes produced.

TABLE II.—Continued.

With Borax on Platinum Wire.	*With Salt of Phosphorus on Platinum Wire.*
O. F.: Dissolved to a limpid glass, which, when the quantity of vanadium oxide is small, appears colorless, when larger, yellow, and which, on cooling, becomes greenish-yellow. R. F.: The glass, while hot, appears brownish, and assumes a fine chrome-green color on cooling.	O. F.: Soluble to a limpid glass, which, if sufficient vanadium oxide be present, appears dark-yellow while hot, and becomes light-yellow on cooling. R. F.: As with borax.
O. F.: Dissolves readily, and in large quantity, to a limpid glass, which appears yellowish while hot; on cooling, it is colorless. When much of the oxide is present, the glass may be made enamel-white by flaming; and on a still larger addition it becomes enamel-white on cooling. R. F.: The saturated glass becomes at first gray and cloudy, and finally transparent again. On charcoal the oxide becomes reduced, the metal vaporizes and coats the charcoal with oxide.	As with borax.

fact that some minerals occur sometimes with, and sometimes without, metallic lustre, these minerals will generally be found enumerated in both groups.

The same precaution has been taken in regard to those species in which the degree of fusibility, whether below or above 5, may appear doubtful.

In the examination of a mineral it is necessary to begin with the first group, and proceed to the following in the order given, for often a mineral belonging to one group has the characteristics of a following one; but the latter may not show the reactions of the former.

The same care is necessary in observing the distinctions between the various divisions, sections, and species.

These divisions are based upon the following properties, and are determined as follows:

1. **Lustre.** In the group of minerals with metallic lustre are placed only those which are perfectly opaque.

If a thin splinter is held between the eye and the light, or placed upon a piece of white porcelain, and shows no translucency, it is considered as having metallic lustre; otherwise as without it.

Moreover, the mineral should have the appearance of some metal as to lustre, and retain this to some extent at least after it is powdered.

The minerals without metallic lustre yield an earthy powder, as well as some with the metallic, which in other respects might be considered as metallic.

Fusibility. The degree of fusibility may be determined by the following scale:

1. **Stibnite.** Fusible in coarse fragments in the flame of a candle.

2. **Natrolite.** Fusible in fine splinters in the flame of a candle.

19

3. Almandite (Alumina-iron garnet). Ea?
fore the blowpipe, but infusible in the flame

4. Actinolite. Fusible in coarse spli?
blowpipe.

5. Orthoclase. Fusible in fine splinters bel
pipe.

6. Bronzite. Fusible on the points and ?
fine splinters before the blowpipe.

In order to test the fusibility of a mi?
splinter, having a sharp edge or point, sh?
off and held in the forceps at a short distan?
point of the inner blue flame, so that the ?
strongly heated. If a gas flame be employ
eral must be held somewhat farther from t?
blue flame than is necessary in the case of ?
order to prevent any reduction taking place,
materially interfere with the results. If a ?
stance is to be tested, or one which decre
heated, and which must therefore be previ
ized, the following process may be resorted
quantity of the powder is made into a pas
and spread upon a piece of charcoal; it i
and strongly heated with an oxidizing flame
(generally) cohere sufficiently to allow of it?
up between the forceps and tested in the u?
Care must be taken that the substance, if ?
and one which acts upon platinum, does n
the platinum points of the forceps.

The fusibility, when the same as actinol
nated by 4; when between that of natrolite
dite, by 2.5; and so on.

A list of oxidized minerals, arranged ?
their fusibility and behavior with sodium ?

fore the blowpipe, may be found at the end of this chapter, page 291.

Hardness. The scale of hardness, as introduced by Mohs, consists of the following minerals:

1. **Talc**: Common laminated, light-green variety.
2. **Gypsum**: Crystalline variety.
3. **Calcite**: Transparent variety.
4. **Fluorite**: Crystalline variety.
5. **Apatite**: Transparent variety.
6. **Orthoclase**: White cleavable variety.
7. **Quartz**; Transparent.
8. **Topaz**: Transparent.
9. **Corundum**: Cleavable varieties.
10. **Diamond.**

The hardness of a mineral may be determined by attempting to scratch it with the minerals enumerated in the scale, or by abrasion with a file. If the file abrades the mineral under trial with the same ease as No, 4, and produces an equal depth of abrasion with the same force, its hardness is said to be 4. If with more facility than 4, but less than 5, the hardness may be 4.2 or 4.5. Several trials should be made to obtain accurate results; and, when practicable, both methods should be employed.

In case a set of minerals like the above is not at hand we may estimate the hardness as follows:

1. Yields easily to the finger-nail.
2. Yields with difficulty to the nail; does not scratch a copper coin.
3. Scratches a copper coin; is also scratched by it, being of about the same degree of hardness.
4. Not scratched by a copper coin; does not scratch glass.
5. Scratches glass, though with difficulty, leaving its powder upon it. Yields readily to the knife.

6. Scratches glass easily. Yield
knife.

7. Does not yield to the knife?
a file, though with difficulty.

8, 9, 10. Harder than flint.

Color. The color of metallic
exposure to the air and light, and
therefore be examined. With non-
color varies greatly.

Streak. This is determined by
upon a piece of unglazed porcela
color of the mark produced. If the
it may be tested by scratching with
streak of metallic minerals is gene
color of the specimen, and the stre
lic is lighter.

Specific gravity. The ordin
can be used for its determination,
described on page 30, will be foun
nient for this purpose.

Crystallization and **cleavage**
aid largely in determining the spec

The systems of crystallization are

I. **Isometric.** The three axes eq
tangular in intersections.

II. **Dimetric** or **Tetragonal.** The
and unequal to the vertical; rectan

III. **Trimetric** or **Orthorhombic.**
equal and rectangular in intersect

IV. **Monoclinic.** The three axes
oblique inclination made by the int

V. **Triclinic.** The three axes une
inclined to one another.

VI. Hexagonal. The vertical axis at right angles to the three lateral, which intersect at angles of 60°.

Water and formation of **Hepar** may be determined by methods already given (pars. 9 and 11).

For most hydrates the temperature of a good gas flame is sufficiently high, but for some silicates, especially the magnesium, the strongest blast flame is necessary, with long-continued heating.

Decomposition by acids. For this test the mineral should be ground as fine as possible in an agate mortar, and placed in some vessel in which the action of the acid and color of the solution can be distinctly seen, and boiled for a quarter of an hour if necessary. If there is no apparent change in the powder, the liquid should be decanted or filtered, and then tested with ammonium carbonate in excess and a few drops of sodium phosphate. If no precipitate or cloudiness is formed, the mineral has not been decomposed.

Formation of a jelly. If silicates finely powdered are heated with phosphoric acid until the acid begins to fume, then cooled, water poured over the substance and evaporated by boiling, the silica separates in form of a jelly. Many silicates gelatinize after ignition, as, for example, garnet, vesuvian, etc. Some splinters or small fragments of the assay are fused or ignited, finely powdered, and boiled in a test-tube with dilute acid. On evaporation, lumps of jelly may be seen, or, if left standing for some time (twelve hours), a stiff jelly may be formed. If water is added, stirred with a glass rod, and then filtered, the solution may be tested for bases with ammonium hydrate, ammonium oxalate, etc., for alumina, lime, etc.

Pyro-electricity. Electricity is developed in some

19 *

minerals by heat. They may be tested with a
or with fibres of wool or cotton.

When the mineral is transparent, a Nicol pi
Von Kobell stauroscope will be needed for
its optical properties.

It is scarcely necessary to add that only pi
mogeneous material will give definite and
reactions.   If the material is found to be imj
must be had to this fact, and conclusions m:
ingly.   Many specimens of wollastonite effe
acids—a property which does not belong to t
but comes from the calcite mixed with it.

Well-known species, with marked characteri;
be studied until some practice is acquired, ar
ers requiring more careful observation and st
examined.

SYNOPSIS OF TABLE!

GROUP I. MINERALS WITH METALLIC

(Of those minerals whose lustre may be uncertain,
placed in this group as are perfectly opaq

CLASS I. Native malleable metals and mercu
CLASS II. Fusibility 1-5, or readily volatile

Division 1.   Before the blowpipe on charc
strong garlic odor of *arsenic*,   .   .

Division 2.   Before the blowpipe on charcc
open tube, give the horse-radish odor of *seleni*

Division 3.   Before the blowpipe on chai
white coating, and color the reducing flam
presence of selenium, greenish-blue.   If *gent*
a small test-tube with an excess of concentrat

acid, it colors the acid hyacinth-red ; but if water is added the color disappears, and grayish-black *tellurium* is precipitated, ... p. 230

Division 4. Before the blowpipe on charcoal give *antimony* fumes, ... p. 231

Division 5. Before the blowpipe on charcoal give with soda a *sulphur* reaction, or, heated in an open glass tube, give off sulphur dioxide, but do not give reactions of the preceding divisions, ... p. 233

Division 6. Do not belong to the preceding divisions, ... p. 235

CLASS III. Infusible, or fusibility above 5, and not volatile.

Division 1. Before the blowpipe give to the borax bead, in very small quantities in the oxidizing flame, an amethyst color (*manganese*), ... p. 236

Division 2. Magnetic, or before the blowpipe on charcoal become so, if strongly heated in the reducing flame, ... p. 237

Division 3. Not belonging to the preceding divisions, ... p. 238

GROUP II. MINERALS WITHOUT METALLIC LUSTRE.

CLASS I. Easily volatile, or combustible, . p. 239

CLASS II. Fusibility 1-5; not, or only partially, volatile, ... p. 240

PART I. Before the blowpipe with soda on charcoal give a metallic globule, or, fused alone in reducing flame, a magnetic metallic mass.

Division 1. Before the blowpipe give with soda a globule of *silver*, ... p. 240

Division 2.   Before the blowpipe give ν globule of *lead*, ...

Division 3.   Before the blowpipe, moisten drochloric acid, color the flame blue, and give acid a solution which, on addition of an exces nium hydrate, assumes an azure-blue color (*coʲ*

Section 1.   Before the blowpipe on charcoal g *arsenic* odor, ...

Section 2.   Before the blowpipe on charcoal giι odor, ...

Division 4.   Before the blowpipe impart tι bead a sapphire-blue color (*cobalt*),   .

Division 5.   Before the blowpipe fused in on charcoal in reducing flame, give a black c tallic magnetic mass.

Section 1.   Give during fusion on charcoal a stι odor, ...

Section 2.   Soluble in hydrochloric acid, withc perceptible residue and without gelatinizing,   .

Section 3.   With hydrochloric acid gelatinize or with separation of silica, ...

Section 4.   But little affected by hydrochloric a

Division 6.   Not belonging to either of ι ing divisions, ...

PART II. Fused with soda on charcoaι metallic globule, or, fused alone in flame, no magnetic metallic mass.

Division 1.   After fusion and continued charcoal in the forceps or on platinum foil, ι kaline reaction, and change to reddish-browι of moistened turmeric paper. (Fragments, an der, should be used.)   .   .   .   .

CLASS III. Infusible, or fusibility above 5.

Division 3.   After ignition ha`
and change to reddish-brown, the .
meric paper, ...

Division 4.   Completely solub`
drochloric or nitric acid, without g
tion or leaving a considerable resi

Division 5.   With hydrochlori
compose, with separation of silic

Section 1.   Before the blowpipe ir

Section 2.   Before the blowpipe in
or but a trace,   .   .   .   "

Division 6.   Not belonging to .
divisions.

Section 1.   Hardness below 7, .

Section 2.   Hardness 7, or above

GROUP I. MINERALS WITH I
(Of those minerals whose metallic lustr
are perfectly opaque are includ

CLASS I.   Native malleable m

Maldonite, Au₂Bi.   Color, silve
black.   Easily fusible on charco
coating and a gold bead.

Silver, see par. 286, Chapter `
trum (alloy of silver and gold), s

Copper, see par. 218.

Lead, characterized by coating
27) and softness; H. = 1.5.

Platinum, see par. 237.   *Iridosm*

Palladium, distinguished from tl
soluble in nitric acid.   Color steel

Native Iron, see par. 239.

Mercury, see par. 276. *Amalgam* Ag, Hg, and Ag_2Hg_3.

Argentite and *Hessite* are malleable, for which see par. 287 and p. 230.

CLASS II. Fusibility 1-5, or readily volatile.
Division 1. Give a strong arsenic odor on charcoal.

Arsenic, see par. 203.

Dufrenoysite, see par. 170; *Sartorite,* $Pb_2As_2S_4$. Before the blowpipe nearly the same as dufrenoysite, but decrepitates strongly; H. $= 3$. *Jordanite,* $Pb_3As_4S_9$. Color, lead-gray; streak black. *Tennantite,* $Cu_8As_2S_7$. Color, iron-black; streak gray. *Epigenite* $(Cu,Fe)_9As_2S_{13}$, similar in its properties, but is trimetric in crystallization.

Polybasite, see par. 293. **Domeykite,** see par. 222.

Binnite, $Cu_6As_4S_9$. In the closed tube gives a sublimate of arsenic sulphide; in the open tube, a crystalline sublimate of arsenic trioxide, with sulphur dioxide. Before the blowpipe on coal gives a faint white coating and odor of arsenic; with soda fuses to a globule, giving metallic copper. Lustre metallic; color black on fresh fracture; streak cherry-red; brittle. H. $= 4.5$; G. $= 4.4$.

Enargite, $Cu_3As_2S_4$. In the closed tube decrepitates and gives a sublimate of sulphur; in the open tube gives off sulphur dioxide and arsenic trioxide, the latter condensing to a sublimate containing often antimony trioxide. The roasted mineral gives a globule of copper with fluxes. Lustre metallic; color and streak grayish-black; brittle; fracture uneven. H. $= 3$; G. $= 4.4$.

Rionite $(Cu,Fe)_6(As,Bi)_2S_8$. With sulphur and potassium iodide gives the red bismuth sublimate.

Algodonite, Cu_6As. H. $= 4$; G. $= 7.6$. See par. 222.

Whitneyite, Cu_9As. Less fusible than al
wise as in domeykite. Massive; cryst
granular. Lustre dull, but strong metall
ed; soon tarnishing. Color bronze to re
coming brown and black on exposure. M
3.5; G. = 8.3.

Smaltite, see par. 214; ~~Cobaltine~~
Skutterudite, Co,As_3; ~~Glaucodot (Co,~~
site $(Co,Fe,Zn)_4(As,Bi)_7S_6$. Before the bl
sapphire-blue color to the borax bead. I
nitric acid, with separation of arsenic a
a red solution. In a concentrated soluti
water gives a cloudiness, but not in the e
skutterudite, and glaucodot, heated in a
sublimate of metallic arsenic. Cobal
The strong acid and dilute solutions a
glaucodot give a precipitate with bariu
solutions of smaltite and skutterudite,
small one. Smaltite has octahedral clea
ite, cubical.

Some varieties contain nickel and i
thite, in which case the nitric acid so
The nickel varieties are distinguished
the powdered mineral with a small am
trated nitric acid, carefully neutralizin
without filtering, and afterward filtering
The filtrate will have a fine blue color.

Glaucopyrite, Fe, Co, Cu, Sb, As, S.
ilar to glaucodot. H. = 4.5; G. = 7.1

Compare the following minerals, also
which often contain cobalt as an imp

Niccolite, see par. 280; *Gersdorffite,*
anthite or *Chathamite* (var. of *Smaltit*

distinguished from gersdorffite by not giving the reactions for sulphur. The proportions of Fe, Ni, and Co vary greatly.

Rammelsbergite, NiAs₂ (similar to chloanthite). In closed tube gives a sublimate of metallic arsenic. See also *Corynite,* Ni(As,Sb)S, which before the blowpipe, on coal, gives the smell of arsenic and fumes of antimony. Color silver-white to steel-gray. *Wolfachite,* Ni(As,Sb)S. Similar in composition, but trimetric in form. See *Ull-mannite,* page 282; *Arsenopyrite,* see par. 243.

Löllingite, FeAs₂. Fuses only on the surface, and with difficulty after the arsenic is driven off. G. = 6.8–8.7.

Compare also bismuth and antimony, which often contain arsenic, but are easily recognized by the white or yellow coating on coal. *Proustite* and *Pyragyrite* often have metallic lustre, but are recognized by their red streak. See pars. 290, 291. *Geocronite* also contains arsenic, see par. 256.

Division 2. Before the blowpipe on charcoal, or in an open tube, give the horse-radish odor of selenium.

Tiemannite, HgSe, *Lehrbachite* (Hg,Pb)Se. Mercury and lead selenide yield metallic mercury on being heated with soda in a closed glass tube (par. 105); the latter yields a globule of metallic lead on being heated on charcoal with soda.

Gaudalcazarite (Hg,Zn)(S,Se). General properties like the preceding, but gives the sulphur reaction.

Clausthalite, PbSe. Color lead-gray; volatilizes without previous fusion, depositing first a slight gray, then a white, and finally a yellowish-green coating; with soda yields with difficulty globules of lead.

20

Naumannite $(Ag,Pb)_5Se_3$.   Color i
readily, and yields with borax a globu

Berzelianite, Cu_2Se, and *Eucairite* (C
of the former silver-white; of the latter
tinguished from the foregoing mineral
by giving copper reactions.

Crookesite $(Cu_2Tl,Ag)Se$, is similar,
per cent. of thallium, and coloring t
green.   *Zorgite* $(Pb,Cu_2)Se$, colors the

**Division 3. Before the blowpipe or
a white coating and color the :
green; in presence of selenium,**

The assay-piece used for this experit
be very small.   It must be borne in m;
erals of this division frequently evolve
nium, owing to a small percentage of
they contain as adventitious constituen

The minerals of this division may b
cording to their color.   .

a. Ores of tellurium, of *tin-white, sil*
dish-white color.

Native Tellurium fuses readily and is

Tetradymite, see par. 209. *Sylvanite,* see par. 236.

Joseite $Bi_{n}Te_{p}SeS_{r}.$

Nagyagite $(Pb,Au)(Te,S)_{2}.$ Color, blackish lead-gray. Distinguished from the preceding by its solution in nitric acid giving a copious precipitate with sulphuric acid. **Compare** aikenite.

Division 4. Before the blowpipe on charcoal give antimony fumes (see par. 16) with a pure white coating.

The fumes possess sometimes the odor of sulphur dioxide or arsenic.

Antimony, distinguished by its tin-white color. Before the blowpipe it takes fire and burns without blowing, and becomes covered with white needles of antimony oxide. **Stibnite,** see par. 200; *Zinkenite,* see par. 256; **Jamesonite,** see par. 256; **Bournonite,** see par. 256.

The powdered stibnite, on being treated with caustic potash, assumes a yellow color, and is for the most part dissolved, while the latter three minerals, which are steel- or lead-gray, do not change color. Bournonite, on being treated with nitric acid, imparts to the solution a sky-blue color. This solution, with sulphuric acid, gives a white precipitate of lead sulphate, and with an excess of ammonia becomes violet-blue. *Stylotypite* $(Cu_{2}Ag_{2}Fe)_{3}Sb_{2}S_{6},$ is similar to bournonite, but no precipitate is formed with sulphuric acid from its solution in aqua regia. Zinkenite and jamesonite are converted into white powders by treatment with nitric acid, without imparting a color to the acid; they are distinguished by their hardness, that of zinkenite being 3.5, that of jamesonite, 2.5. The former has no cleavage, while in the latter it is very marked in one direction.

Ullmannite, see par. 282; **Berthierite**, see par. 201; *Breithauptite*, Ni,Sb. All yield a magnetic globule with continued heat. Breithauptite is distinguished from the other two by not giving a sulphur reaction. Color bright copper-red.

Division 5. Before the blowpipe on charcoal give with soda a sulphur reaction, but do not give the reactions of the preceding divisions.

Argentite, see par. 287; *Jalpaite*, see par. 287; *Acanthite*, see par. 287.

Galenite, see par. 255.

Cinnabar, see par. 277. *Metacinnabarite* is amorphous HgS; streak black; H. = 3; G. = 7.72.

Alabandite, MnS. Isometric. H. = 3.5. Color iron-black; powder leek-green; lustre submetallic.

Hauerite, MnS_2. H. = 4. Color brownish-black; powder brownish-red; lustre metallic adamantine. Yields sulphur on being heated in a matrass. These manganese minerals, boiled with a mixture of phosphoric and nitric acids, give a fine violet solution. In the borax bead give also a violet color.

Chalcocite, see par. 219; *Stromeyerite*, see par. 288; **Stannite**, see par. 298; **Chalcopyrite**, see par. 220; **Bornite**, see par. 221; *Cubanite*, $Cu_2Fe_2S_4$; *Wittichenite*, Cu_3BiS_3; *Emplectite*, $CuBiS_2$, color tin-white; *Aikinite*, $CuPbBiS_3$; *Grünauite*, Ni,Bi,Fe,Cu,S; *Cuproplumbite*, $Cu_2S,2PbS$; *Pentlandite* (Ni,Fe)S. All these minerals are partially soluble in nitric acid, the solution possessing a sky-blue or green color; on addition of water to the concentrated solution, a white precipitate is produced if the mineral under examination were wittichenite, grünauite, or aikinite. [To distinguish these three, add to

20*

the acid solution sulphuric acid: a prec
aikinite; wittichenite gives the copper r
treated as described in par. 90 ; grünau
copyrite and cubanite are distinguished
by their brass-yellow color; bornite is al
by its color. To distinguish the remainii
make a solution in nitric acid; add sulph
cipitate indicates cuproplumbite; if no p
duced, add hydrochloric acid : a precipita
meyerite; to distinguish between chalcoc
see pars. 219 and 298. *Castillite* resemb

Huascolite, a zinciferous variety of g

Millerite, see par. 279; **Linnæite**

Pyrite, see par. 240; **Marcasite**, see

rhotite, see par. 242; *Sternbergite*, AgF
bers of this subdivision fuse to globules wt
by the magnet. They are readily disti
characteristics given in Chapter VII.
the treatment described in par. 117, yie
silver. Marcasite, trimetric, and pyrite
only be distinguished by their crystalli

Carrollite, Co$_2$CuS$_4$. Tin-white color.
ilar to linnæite, but also, when moister
chloric acid, gives a blue color to the

Beyrichite, Ni$_5$S$_7$. Color lead-gray.
millerite, but gives sulphur in the clos

Bismuthinite, see par. 208; **Chivi**
Bi$_6$S$_{11}$. Decomposed with nitric acid, wi
lead sulphate.

Bismite, Bi$_2$O$_3$. Pulverulent or earth;
ish-yellow to white.

Division 6. Do not belong to the preceding divisions.

Amalgam, see par. 276; *Arquerite,* see par. 276.

Bismuth, see par. 207.

Rabdionite, $(Cu,Mn,Co)(MnFe)O_4$. Stalactitic; color black; streak metallic gray. Colors the borax bead blue; heated with phosphoric acid, colors the solution violet.

Hematite, see par. 244.

Cuprite, see par. 227. Often with weak metallic lustre.

Magnetite, see par. 246.

Hortonolite $(Fe,Mg)2SiO_4$, yellowish-black, and **Fayalite,** Fe_2SiO_4, black, and both magnetic before ignition, and gelatinize in hydrochloric acid.

Wolframite $(Mn,Fe)WO_4$. H. $= 5$-5.5; G. $= 7.1$-7.5. Monoclinic. Lustre submetallic; streak dark reddish-brown to black; opaque; sometimes magnetic; color dark-grayish or brownish-black; fusibility 3. The pulverized mineral, on being boiled with aqua regia, is decomposed and assumes gradually a yellowish color. Boiled for some time with phosphoric acid, gives a fine blue syrup, especially after cooling (*tungsten*). If diluted with water it becomes reddish-yellow, and finally colorless. On the addition of iron filings and sulphuric acid, and shaking, it gradually becomes sapphire-blue. This solution, diluted with water, loses its color again after a little time. If the blue syrup is treated with phosphoric and nitric acids it becomes violet (*manganese*).

The variety *dianite,* when treated as above, gives, when boiled with tin foil and concentrated hydrochloric acid, and diluting with its volume of water, a sapphire-blue fluid; while with the tantalite and ordinary columbite the metallic acids remain undissolved and the filtrate colorless.

The color of this mineral is iron-blac
yttrotantalite, grayish; of dianite, gr
reddish-brown; of columbite, brownisl
lite, brown.

Samarskite, $Cb_2O_5, Ta_2O_5, WO_3, SnO_2$
$MnO, FeO, CeO, YO, CaO, H_2O.$ Color
tre of surface of fracture shining and su
dark reddish-brown; fusibility 4.5. B
verized mineral with potassium hydrat
cible, boiling the fused mass in water,
solution, which is filtered; hydrochlo
white precipitate. If this is boiled v
hydrochloric acid and tin for a few n
volume of water added, it gives a bri

Rhodonite, dark varieties, MnSiC
upon by acids. In a fine powder turn
chloric acid, with slight separation of si
flame, colors the borax glass amethyst.

Some varieties of *Psilomelane*, see pa

Fayalite, Ilvaite, and **Allanite**, s
Div. 5, p. 249.

Plattnerite, PbO_2. Color iron-black
adamantine; streak brown; opaque;
metallic lead with soda. $G. = 9.3.$

CLASS III. Infusible, or fusibility a volatile.

Division 1. Before the blowpij
borax bead, in very small qu
oxidizing flame, an amethyst co

The members of this division are d
each other principally by their physic

The manganese oxides are more or

in hydrochloric acid, with evolution of chlorine. If powdered and boiled with phosphoric acid to a syrupy consistency, the solution becomes a fine violet color, which,
diluted with water and shaken with some crystals of ferrous sulphate, becomes colorless.

Compare franklinite, in the next division, which is also magnetic.

Lithiophorite, $MnO_2, Al_2O_3, LiO, H_2O$, colors the flame
carmine-red.

Braunite, see par. 270; *Hausmannite,* see par. 269;
Psilomelane, see par. 273; **Pyrolusite,** see par. 268;
Franklinite, some varieties, see par. 247; **Manganite,**
H_2MnO_4. Color steel-gray to iron-black; streak dark
reddish-brown. $H. = 4$. Yields much water in a ma-
trass. Otherwise like braunite.

Crednerite, $Cu_3Mn_2O_9$. $H. = 4.5$. Lustre metallic;
color iron-black to steel-gray; streak brownish-black.
Moistened with hydrochloric acid, gives a fine blue color
to the flame. Dissolved in hydrochloric acid with an excess of ammonia, a precipitate is formed and a blue solution, which is not the case with the foregoing.

Compare *Alabandite* and *Hauerite.*

Division 2. Magnetic, or, heated on charcoal in the reducing flame, become magnetic.

Löllingite and *Arsenopyrite,* some varieties are infusible, but may be distinguished by the arsenic odor given
off on charcoal before the blowpipe.

Hematite, see par. 244; **Turgite,** see par. 250.
Franklinite, see par. 247. **Magnetite,** see par. 246.

Jacobsite $(Mn,Mg)(Fe,Mn)O_4$; *Magnesioferrite,* Mg
FeO_4. Color and streak of both minerals black; more
or less magnetic without heating; dissolve with difficulty
in hydrochloric acid. After oxidation of the solution

with potassium chlorate, and precipitation with an excess of ammonia, sodium phosphate precipitates the magnesia. Jacobsite gives the manganese reactions.

Menaccanite, see par. 245.

Compare *Rutile*, *Arkansite*, and *Anatase*, which are often magnetic from the presence of titanic iron, or become so after continued ignition. They are scarcely attacked by hydrochloric acid.

Some varieties of **Limonite,** see par. 248, *Siderite*, and *Sphalerite*, see par. 300, have metallic lustre. See *Graphite*.

Division 3. Not belonging to the preceding divisions.

Chromite, see par. 212. *Cassiterite* often has a similar metallic lustre. Easily reduced on charcoal with potassium cyanide.

Molybdenite, MoS_2; **Graphite,** C., see par. 305. Both very soft; hardness 1.5. Molybdenite, when heated in the forceps, colors the flame greenish, and gives a sulphur reaction when treated as described in par. 121. Color bluish-gray; sectile and nearly malleable. Decomposed by nitric acid, leaving a white or grayish residue of molybdenum trioxide.

Perofskite, $CaTiO_3$. Isometric. Gives the reaction for titanic acid as described in par. 125. Distinguished by crystalline form.

Compare *Rutile* and *Brookite*, Div. 6, p. 287.

Iridosmine, see par. 238.

Tantalite, $Fe(Mn)Ta_2O_6$, and *Columbite*, $FeCb_2(Ta_2)O_6$; *Yttrotantalite* $(Fe,Ca,Y)_2(TaCb)_2O_7$. The color of these minerals is iron-black; yttrotantalite loses its color before the blowpipe and becomes yellowish or white; that of the others remains unchanged. Acids affect them but little. If tantalite and columbite are powdered, fused with caustic potash in a silver crucible, dissolved in water, and filtered, a precipitate is formed with hydrochloric acid, which, boiled with dilute sulphuric acid, becomes white; on the addition of zinc the precipitate from the columbite becomes intense blue in the hot solution, and retains this color on the addition of water for a considerable time. The precipitate from tantalite is lighter colored, and loses its color quicker with water.

Compare *Polycrase*, Div. 4, p. 280, and *Æschynite*, Div. 6, p. 287.

Uraninite, U_3O_8. Color usually velvet-black; lustre greasy; partially soluble in nitric acid to a yellow liquid; the solution gives a sulphur-yellow precipitate with ammonia. Boiled with phosphoric acid gives an emerald-green solution. G. = 6.4–7.

GROUP II. MINERALS WITHOUT METALLIC LUSTRE.
CLASS I. Easily volatile, or combustible.

Native Sulphur, S. H. = 1.5–2.5; G. = 2. Completely volatile; burns with a blue flame and emission of sulphur dioxide. Color sulphur-yellow, honey-yellow, and gray or brown from impurities.

Realgar, see par. 205; **Orpiment,** see par. 204.

Arsenolite, see par. 206; **Kermesite,** see par. 202.

Valentinite, Sb_2O_3. Trimetric. Color white; streak white; lustre adamantine; does not change color with potassium hydrate; does not evolve sulphuretted hydrogen with hydrochloric acid, but dissolves easily.

Senarmontite, Sb_2O_3. Isometric. Lustre resinous, inclining to sub-adamantine; streak white.

Sal-ammoniac, NH_4Cl; *Mascagnite* $(NH_4)_2SO_4$ + aq. Color white. Both evolve ammonia with potassium hydrate; the former is volatile without previous fusion; the latter intumesces. It also gives a precipitate with barium chloride.

Cinnabar, see par. 277; *Calome*

Cotunnite, PbCl$_2$.   H. = 2; G
streak yellowish-white.   Fuses easi
and gives a white coating, the in
tinged yellow; with soda on charc
metallic lead.

CLASS II. Fusibility 1-5; not
volatile.

PART I. Give with soda on
lic globule, or, fused alone
a magnetic metallic mass.

(All minerals without metallic lustre, wh
belong to this group, excepting

Division 1. Give with soda a
(It is well to fuse the globule with borax
pure and malleabl

Proustite, see par. 291; **Pyra**
Xanthoconite, Ag$_9$As$_3$S$_{10}$, behaves
which it is distinguished by its or;
streak.

Compare *Myargyrite*, Div. 4, p. 232. wh
Pyrargyrite.   The G. of the former is 5.2

Cerargyrite, see par. 294; *Iody*
bolite, see par. 296.   All malleable

Division 2. Give with soda

The minerals of this division a;
acid; the solution gives a copious
phuric acid.   If dissolved by boilin
potassium chromate directly, or on ;
gives an orange precipitate.

Bindheimite, $Pb_3Sb_2O_8 + 4aq$, and *Nadorite*, PbSb ClO_2. Lustre resinous or dull; color and streak white, grayish-yellow. Before the blowpipe on coal give a coating of lead and antimony. The former in the matrass, water. The latter, fused in a salt of phosphorus bead with copper oxide, colors the flame blue (*copper chloride*).

Mimetite, $3Pb_3As_2O_8 + PbCl_2$; *Hedyphane,* $3(Pb, Ca)_3As_2O_8 + (PbCa)Cl_2$. Before the blowpipe on charcoal, the former completely, the latter partially, reduced to metallic lead, with evolution of arsenic fumes. If mimetite is fused, on cooling it crystallizes like pyromorphite.

Pyromorphite, see par. 263.

Minium, see par. 257; **Crocoite,** see par. 260; *Phœnicochroite*, $Pb_3Cr_2O_9$; *Dechenite* $(Pb,Zn)V_2O_6$. Crocoite and phœnicochroite give the chromium reaction (par. 84). Dechenite gives to the borax bead an emerald-green color, which becomes light olive-green in the oxidizing flame, then yellow and colorless. These three, on being boiled with a large amount of hydrochloric acid and for a long time, give an emerald-green solution, with separation of lead chloride; on adding alcohol to the liquid, concentrating by heat, pouring off from the residue, and then adding water, the liquid assumes a sky-blue color if dechenite is present; if the other minerals, green. The streak of crocoite and dechenite is reddish-yellow, and that of phœnicochroite brick-red.

Linarite, $PbCuSO_5 + aq$, is characterized by its deep azure-blue color. The color is destroyed by digesting with nitric acid, and lead sulphate is precipitated. In the closed tube gives off water.

Cerussite, see par. 265; *Phosgenite,* see par. 267;

is of similar composition,
PbCO, + PbSO₄. ...All no
rescence; leadhillite and li
residue. The solution of
nitrate, a precipitate of si

Mendipite, **Pb₃O₂Cl₄.**
age; colorless, white. *M*
cleavage; green to yellow
acid *without effervescence,*
itate with solution of silv

Anglesite, PbSO₄, see
Wulfenite, PbMoO₄, s
Stolzite, PbWO₄. Cole
red; lustre resinous. Sol
hydrochloric acid, leaving
WO₃. With sulphuric ac
sumes a bright lemon-ye
colored.

Vauquelinite, Pb₂CuCr₂C
3Pb₃V₂O₈ + PbCl₂. Hexa
blackish to olive-green; c
ish. Both impart to the
color; both are soluble in
vanadinite is yellow, and
nitrate; that of vauquelini
same as vanadinite, but t

Compare *Plumbogummite,* pa

Laxmannite, (Pb,Cu)₆(I
green; lustre vitreous. D
nium molybdate gives a
acid).

Division 3. When moistened with hydrochloric acid, color the flame blue, and give with nitric acid a solution which, on addition of an excess of ammonia, becomes azure-blue.

The copper oxide minerals of this group are for the most part decomposed to such an extent by boiling with caustic potash that their acids combine with the potash.

Section 1. Before the blowpipe on charcoal give a strong *arsenic* odor, and most yield a white brittle globule of copper arsenide. They are of green color.

Chenevixite, $(Fe, Cu_3)_2As_2O_{11} + 3aq.$ Lustre vitreous; color dark-green; streak yellowish-green. Fuses to a black magnetic slag, while the following do not:

Bayldonite, $(CuPb)_4As_2O_9 + 2aq.$ Lustre resinous; color green. Dissolved in nitric acid, gives a precipitate with sulphuric acid of lead sulphate. In a closed tube gives off water and becomes black.

Olivenite, see par. 230; *Clinoclasite*, $Cu_6As_2O_{11} + 3aq.$ Color dark bluish-green. In matrass gives 7 per cent. of water, and olivenite 4 per cent.

Tyrolite, see par. 231; *Chalcophyllite*, $Cu_8As_2O_{13} + 12aq.$ Color emerald- to grass-green. Both decrepitate violently and yield much water; chalcophyllite dissolves in ammonia without leaving a residue. H. = 2.

Conichalcite, $2(Cu, Ca)_4(As, P, V)_2O_9 + 3aq.$ Color pistachio- to emerald-green. H. = 4.5. Fused, gives an alkaline reaction.

Liroconite; $(Cu_3Al_2)(As, P)_2O_{11} + 12aq.$ Color sky-blue to green; does not decrepitate in the matrass; changes to a smalt-blue color when gently heated; loses 22 per cent.

of water on ignition. Soluble in ammonia, with white flocky residue.

Euchroite, $Cu_4As_2O_9 + 7aq$; *Erinite*, $Cu_5As_2O_{10} + 3aq$. Color of both emerald-green. The former loses by ignition 19 per cent. of water; the latter only 5 per cent. Erinite amorphous. *Cornwallite*, $Cu_5As_2O_{10} + 3aq$, also amorphous. Loses 13 per cent. of water on ignition.

Section 2. Before the blowpipe on charcoal give no *arsenic* odor, but most yield a malleable copper bead.

Atacamite, see par. 226; *Tallingite*, $CuCl_2 + 4H_2Cu$ $O_2 + 4aq$, and *Percylite*, $(Pb,Cu)(Cl,O) + aq$; *Nantokite*, $CuCl$. White; yields no water in a closed tube. **Chalcanthite,** see par. 229; *Brochantite*, $Cu_4SO_7 + 3aq$; **Covellite**, CuS. These three minerals give a sulphur reaction (par. 121); chalcanthite is soluble in water; the other two not. Color of covellite dark indigo-blue; of brochanthite, emerald- to blackish-green, with 12 per cent. of water. *Langite*, $Cu_3SO_6 + 4aq$; greenish-blue color, with 16 per cent. of water.

Cuprite, see par. 227; **Melaconite,** see par. 228. Both dissolve readily in acids without effervescence (except impure varieties of melaconite).

Malachite, see par. 232; **Azurite,** see par. 233; *Mysorin*, $CuCO_3$. Color blackish-brown, usually green or red from mixture with malachite or iron oxide; does not yield water in a matrass. All three dissolve readily in acids, with effervescence. *Aurichalcite*, $(Zn,Cu)_3CO_5 + 2aq$, gives a zinc coating on coal. Color bluish-green. *Atlasite*, $7Cu_2CO_4 + CuCl_2 + 10aq$, dissolved in nitric acid, gives a precipitate with silver nitrate. Celandine- to emerald-green.

Pseudomalachite, $Cu_6P_2O_{11} + 3aq$ (*Lunnite* and *Ehlite*), Libethenite, $Cu_4P_2O_9 + H_2O$. Dark olive-green. Tagilite, $Cu_4P_2O_9 + 3aq$. Verdigris- to emerald-green. Are all readily soluble in nitric acid without effervescence; the (slightly acid) solution gives a precipitate with lead acetate. Pseudomalachite loses 14 per cent. of water on ignition; the others less (from 7 to 10).

Torbernite, $CuU_2P_2O_{12} + 8aq$. Color grass-, leek-, to emerald-green. Dissolves in nitric acid to a yellowish-green liquid, on addition of ammonia in excess, a bluish-green precipitate is formed, the supernatant liquid being blue. Warmed with ammonium molybdate, gives a yellow precipitate.

Volborthite, $(Cu,Ca)_4V_2O_9 + H_2O$. Olive-green to lemon-yellow. Pearly lustre; fuses easily. H. $= 3-3.5$.

Division 4. Before the blowpipe impart to the borax bead a sapphire-blue color (*cobalt*).

Erythrite, see par. 217; *Annabergite*, see par. 283; Heterogenite, $CoO + 2Co_2O_3 + 6aq$. Color black to reddish-brown; difficultly fusible. Colors the flame green.

Division 5. Fused in forceps, or on charcoal in reducing flame, give a black or gray metallic magnetic mass, but do not give the reactions of the preceding divisions.

To observe well the magnetic character of the fused mineral it is advisable to expose a pretty large assay-piece for some time to the action of the reduction flame.

Section 1. Evolve a strong arsenic odor during fusion.

Scorodite, see par. 253; *Pitticite*, $Fe_2O_3, As_2O_5, SO_3 +$

21*

H_2O; *Pharmacosiderite*, $Fe_4As_3O_{17}$ +
streak green, brown, yellow. $H. = 2$
to subtranslucent; somewhat metallic;
dantite, $Fe_2As_2O_{17}$ + 15 aq. The gel
sume with potassium hydrate a reddish
odite and beudantite occur crystallize
and the second rhombohedral. Their
shade of green to brown and black. {
to yellow; lustre vitreous. Pitticite, m
$H. = 2$–3. Lustre vitreous, sometime
lowish, brownish, blood-red, and wl
white; translucent-opaque.

Arseniosiderite, $(Ca_3Fe)As_4O_8$ + H_4
lowish-brown; fibrous; lustre silky.

Morenosite, $NiSO_4$ + 7 aq. Partly
The solution assumes a blue color on
nia; sometimes contains arsenic.

Section 2. Soluble in hydrochlo
  leaving a perceptible residue, a
  tinizing. Give no arsenic odo:
  coal.

Ludwigite, Mg,B,Fe,FeO_8. Finely
green to black. Warmed with sulphu
with alcohol, burns with a green flame

Rabdionite, $(Cu,Mn,Co)(FeMn)O_4$.
yields 13 per cent. of water and colc
cobalt-blue. Color black. Solution
violet.

Stibioferrite, Fe_2O_3,Sb_2O_5 + aq. f
on stibnite; color yellow; gives on
fumes.

Pettkoite, $(Fe_3Fe)S_3O_{12}$, with little (
2.5. Isometric; lustre bright; color

dirty-greenish. Soluble in water; precipitate with barium chloride.

Melanterite, see par. 251; *Botryogen*, $(Fe,Mg)FeS_4$ $O_{16} + 12aq.$ Melanterite and botryogen are soluble in water, the latter leaving a yellow residue. The solutions give precipitates with barium chloride; also with ammonia. Streak and color of melanterite are green; of botryogen, color ochre-yellow to red; streak yellow. *Ræmerite*, yellowish-brown. Coquimbite, *Jarosite*, and *Fibroferrite*, all yellow. The last, fibrous and silky, behaves similarly to botryogen. Belonging to this section are Copiapite, *Raimondite, Pastreite, Carphosiderite*, all giving a yellow powder, and are *insoluble* in water. *Voltaite* is distinguished from the foregoing by its black or dark-green color, resinous lustre, and octahedral crystallization. All these sulphates, when heated in the closed tube, give much water.

Siderite, see par. 254.

Hureaulite, $(Mn,Fe,H_2)_3 P_2 O_8 + 4aq.$ $H.=5.$ **Triplite** $(Fe,Mn)_3 P_2 O_8 + (Fe,Mn)F_2.$ $H.=4.5-5.$ Fuse readily; moistened with sulphuric acid, give the phosphoric acid reaction (par. 60); with borax, strong manganese reaction; hureaulite yields much water; triplite none, or very little.

Sarcopside, $4(Mn,Fe)_3 P_2 O_8 + H_4 FeO_5$, distinguished from triplite by its color — flesh-red to lavender-blue; streak straw-yellow; lustre silky. $H.=4.$

Triphylite, $(Fe,Mn,Li_2)_3 P_2 O_8$, shows a similar behavior; the manganese reaction is much less decided. On dissolving the mineral in hydrochloric acid, evaporating the solution to dryness, adding alcohol, heating the alcohol to ebullition, and burning the vapor, the flame assumes a purple color. Color greenish-gray to bluish; cleavage

perfect; resinous lustre. ▮▮▮▮▮▮
brown variety.

Diadochite, $Fe_2O_3, P_2O_5, SO_3, H_2O$. ▮
Reniform or stalactitic; lustre resinous t
yellowish-brown; streak uncolored; so
chloric acid; gives precipitate with b
when ignited, gives off sulphuric acid.

Vivianite, see par. 252; *Dufrenite*
$P_2O_{11} + 3$ aq; **Cacoxenite**, $Fe_2P_2O_{11}$ +
ite, $(FeCa_3)_5P_4O_{25} + 15$ aq. Fuse read
with sulphuric acid like the preceding
ganese reaction. Yield much water in a
enite, 33 per cent.; vivianite, 28 per cen
per cent.; dufrenite, 10 per cent. Col
leek-green; of cacoxenite, ochre-yellov
various shades of blue; of borickite,
Beraunite, $FeP_2O_8 +$ aq, is a similar p
color.

Hematite, see par. 244.
Compare *Limonite*.

**Section 3. With hydrochloric acid ge
readily decomposed with separation**

Cronstedite, $(3FeMg)_2SiO_4 + Fe_2SiO_3)$
3.5. Rhombohedral, also amorphous; co
dark leek-green; gelatinizes with hydroc
a closed tube gives off water; fuses w
black glass. *Sideroschisolite* is probabl

Stilpnomelane, $(FeMg)_3(Fe,Al)Si_3O$
3.4. **Chalcodite** is similar in composit
vety coatings of brass-like lustre. The co
erals is black, yellowish, and greenish-br
is greenish-gray.

Voigtite, Ekmannite, and *Euralite* are closely related to the above. The first are mica-like in aspect and structure, and the last is massive and yields 11 per cent. of water. Is decomposed by hydrochloric acid without gelatinizing.

Palagonite, of brownish-yellow color and streak; amorphous; yields water (14 per cent.) and fuses to a black magnetic glass. Some varieties gelatinize; others do not. See also *Jollyte.*

Ilvaite, $H_2Ca_2Fe_4FeSi_4O_{18}$; **Allanite,** $(Ce,La,Di,Fe,Ca)_3(AlFe)Si_3O_{12}$, yield no water, or only a trace; gelatinize with hydrochloric acid; allanite fuses with intumescence to a voluminous brownish or blackish glass; ilvaite intumesces but slightly, decrepitates, and fuses to an iron-black bead. Hardness of each, **5.5–6.**

Fayalite, Fe_2SiO_4, and **Hortonolite,** $(Fe,Mg)_2SiO_4$, are crystalline, cleavable, gelatinize perfectly, of resinous lustre, and H. = **6.5.** Color of fayalite is black, greenish, or brownish-black; easily fusible and magnetic. Hortonolite, decomposed by phosphoric acid, the jelly treated with nitric acid, immediately becomes violet.

Knebelite, $(FeMn)_2SiO_4$, shows the same reaction. Color gray, red, brown to black; easily fusible. H. = **6.5.**

Roepperite, $(Fe,Mn,Zn,Mg)_2SiO_4$. Color dark-green to black, difficultly fusible, and gives with soda a sublimate of zinc oxide.

Pyrosmalite, $(FeMn)Cl_2 + 7(Fe,Mn)SiO_3 + 5aq$, and *Astrophyllite,* $(K,Na)_6(Fe,Mn)_{15}(FeAl)_2(Si,Ti)_{16}O_{46}$, containing titanium, and sometimes zirconium, are decomposed by hydrochloric acid, with separation of silica, without gelatinizing. Fusibility **2–2.5.** Pyrosmalite gives the chlorine reaction (par. 65); astrophyllite not. The hy-

drochloric acid solution of the latter giv
for titanic acid. Both minerals are cleav
rection; the latter often micaceous.

Lepidomelane, K₄Fe₂(AlFe)₃Si₃O... F
vitreous; color dark-green to black, with
leek-green reflection; streak grayish-gree
Easily decomposed by hydrochloric acid,
ica in scaly flakes; easily fusible.

Allochroite, $Ca_3FeSi_3O_{12}$. Some varietie
perfect jelly with hydrochloric acid. I
color green, brown, black; lustre greasy

Gillingite, $FeO_3,FeO,MgO,CaO,SiO_2,F$
FeO_3,MgO,SiO_2,H_2O [a variety of *Serpent*
difficulty; do not gelatinize. The former
phous; the latter brown, fibrous, woody. B
in a matrass.

Some impure varieties of **Limonite,** se

Section 4. But little affected by hydr

Crocidolite, $Na,Mg,Fe.,SiO_2 + H_2O$, an
$2(Na_2,Fe,Ca)SiO_3 + Fe_2Si_3O_9$, are easily fi
with much intumescence and escape of ga
black glass. Color of crocidolite, lavend
green; fibrous; yields water in a matrass;
black and yields no water.

[See also *Hornblende* and *Tourmaline*,
of which become slightly magnetic after fu
Lepidomelane.]

Glauconite, $FeO.MgO,K_2O,Al_2O_3,Si$
earth]. Fusibility 3, without swelling, ar
in the matrass; color celandine-green; hard

Acmite, $(Na_6Fe_3Fe)Si_3O_9$; *Babingtonite*
$SiO_3 + FeSi_3O_9$. Fusibility of the former

ter, **2.5.** Form a black lustrous slag. Acmite cleaves at an angle of 93°. Babingtonite, fused with soda, dissolved in hydrochloric acid, and ammonia added to separate iron, the filtrate with ammonium oxalate, gives a heavy precipitate of lime. Acmite gives no lime precipitate.

Compare *Augite.*

Almandine Garnet [iron garnet], $Fe_3AlSi_3O_{12}$. Fuses quietly at **3**; gelatinizes after fusion; hardness **7–7.5**; color reddish-brown; not very cleavable. See also *Allochroite.*

Wolframite, $(Fe,Mn)WO_4$. Color brownish-black; streak brownish; sub-metallic lustre. Boiled with concentrated phosphoric acid, a blue syrup, which, diluted with water, becomes colorless; if powdered iron is added and then shaken, it gives a fine blue color. Decomposed with aqua regia, with separation of a yellow powder, WO_3.

Megabasite, $Mn,FeWO_4$, behaves in a similar manner. Streak ochre-yellow. *Hübnerite,* Mn,WO_4, same reactions, but contains no iron.

Rhodonite, see par. 275.

Lepidolite, $(K,NaLi)_6Al_4Si_{12}O_{30}$. Color rose-red to gray-white; fracture often micaceous; lustre vitreous. H. = **2.5–3**; fusibility **2–2.5**. Often becomes magnetic, and colors the flame reddish-purple.

Compare *Lepidomelane,* p. 250.

Division 6. Not belonging to either of the preceding divisions.

Molybdite, MoO_3. Color sulphur- or orange-yellow; earthy. H. = **1–2**; fusibility = **1**. Gives with the fluxes the reactions of molybdic acid. Dissolves readily in hydrochloric acid; the solution is colorless, but turns blue on being shaken with tin foil. Fuses on coal, fumes, and

is absorbed. With salt of phosphorus i
flame, gives a bead, when cold, of beau

Eulytite, $Bi_4Si_3O_{12}$. H. = 4.5. Fuses ea
bead; gelatinizes with hydrochloric acid
with soda, yields a globule of metallic
dark-brown to yellow; lustre resinous.

Bismutite, see par. 211; *Pucherite*, Bi
Fuses easily; vitreous, adamantine lustre.
phosphorus gives a green bead in the 1
These three minerals give the bismuth re
tassium iodide and sulphur on charcoal.

Compare *Walpurgite*, p. 256; compare also *A*
vious division; also *Lepidomelane*, p. 250.

**PART II. Fused with soda on chai
metallic globule, or, fused alone
flame, no magnetic metallic mass**

**Division 1. After fusion and conti
on charcoal, in the forceps or
foil, have an alkaline reaction,
the color of moistened turmeric
dish-brown. The test may be mad
ters, and not with the powder.**

Section 1. Easily and completely sol

Nitre, KNO_3; Soda Nitre, $NaNO_3$, d
on burning coals. Fused on platinum w
colors the flame bluish, with a red tint; th
yellow. Platinum chloride gives a precip
lution of nitre.

Natron, $Na_2CO_3 + 10aq$; Trona, 1
The aqueous solution has an alkaline reac
vesces on addition of hydrochloric acid.

former decompose quickly in the air; the latter not. *Thermonatrite*, Na_2CO_3, + aq. H. = 1.5. Effloresces in the air, and behaves like the preceding.

Mirabalite, Na_2SO_4 + 10aq; *Thenardite*, Na_2SO_4; *Aphthitalite*, K_2SO_4: Epsomite, $MgSO_4$ + 7aq; **Kalinite,** $K_2AlS_4O_{16}$ + 24aq. The aqueous solutions of these minerals do not effervesce with acids; give a copious precipitate with barium chloride; the solutions of kalinite and epsomite are precipitated by potassium carbonate (distinguished by reaction with cobalt solution, par. 44). *Kainite*, $K_2MgS_2O_8$ + 6aq, behaves in a similar manner; soluble in water, and a precipitate is formed with silver nitrate. The concentrated solution of aphthitalite gives a precipitate with platinum chloride; mirabalite yields much water; thenardite none; the epsomite contains 50 per cent. of water; *Loeweite*, $2Na_2MgS_2O_8$ + 5aq, and *Kieserite*, $MgSO_4$ + aq, each 14 per cent.; *Bloedite*, $Na_2MgS_2O_8$ + 4aq, 21 per cent.; and kainite, 27 per cent. *Picromerite*, $K_2MgS_2O_8$ + 6aq.

Tachydrite, $CaMg_2Cl_6$ + 12aq. Color yellowish; deliquescent; yields much water in the matrass; colors the flame red. *Carnallite*, $KMgCl_3$ + 6aq. Massive, granular; lustre greasy; color milk-white; often reddish from presence of iron oxide; strongly phosphorescent; yields much water; fuses easily; easily soluble in water; and yields precipitate with platinum chloride.

Halite, NaCl; **Sylvite,** KCl. The aqueous solution gives a copious precipitate with silver nitrate; gives also the reactions for chlorine described in pars. 65, 66. The latter gives a heavy yellow precipitate with platinic chloride, but the former does not.

Borax, $Na_2B_4O_7$ + 10aq, gives the reaction for boric acid (pars. 75, 76).

22

Section 2. Insoluble in water, or ~~~~~~~~

Ulexite, $NaCaB_5O_9 + 5aq.$ Fusibility 1,
the flame yellow; yields much water. Moister
sulphuric acid, the flame changes momentarily
Somewhat soluble in hot water, giving alkali
tion.

Gay-Lussite, $Na_2CO_3 + CaCO_3 + 5aq$; **Wi**
$BaCO_3$; *Staffelite,* $Ca_3P_2O_8 + CaCO_3$. Dissolve
hydrochloric acid with effervescence; the first yi
ter; the latter do not. The solution of the staffel
a precipitate with ammonia (calcium phosphate);
ers not. The solution with ammonium molybdate,
gives a yellow precipitate.

Compare *Strontianite*, which colors the flame crimson.

Anhydrite, $CaSO_4$; **Gypsum,** $CaSO_4 + 2aq$
halite, $Ca_2MgK_2S_4O_{16} + 2aq$; *Glauberite,* Na
Soluble in much hydrochloric acid; in the solu
rium chloride gives a heavy precipitate; gypsu
much water; polyhalite little; the rest none; a
is distinguished by superior hardness, 3.5; poly
distinguished from glauberite by its solution givii
low precipitate with platinum chloride.

Syngenite, $K_2CaS_2O_8 + aq$, which occurs in
similar to polyhalite.

Barite, $BaSO_4$; **Celestite,** $SrSO_4$. Insoluble i
chloric acid; give a sulphur reaction when tr
described in par. 121. Celestite colors the fl
(par. 59); barite, yellowish-green (par. 60).

Fluorite, CaF_2; **Cryolite,** Na_6AlF_{12}; *Phar*
$2HCaAsO_4 + 5aq$. Do not effervesce with aci
give no sulphur reaction. Pharmacolite evolves
odor on charcoal; the other two give fluorine
r. 76). **Fusibility** of fluorite, 3; of cryolite,

tozenite (var. of fluorite), gives odor of antozone on being rubbed. *Chiolite*, Na_3AlF_9, behaves like cryolite; occurs only massive-granular; while cryolite is distinctly crystalline, and cleavable in three directions.

Pachnolite, $Na_2Ca_2AlF_{12}2aq$, yields strongly acid water. Closely related are *Arksutite* and *Chodneffite*, without water; and *Gearksutite*, with water.

Cancrinite, $Na_2AlSi_2O_8$, effervesces with hydrochloric acid, and gelatinizes. In the flame it grows white and opaque, and then melts (2.5), intumesces, and forms a white blebby mass. The easy fusibility distinguishes it from nephelite, which, laid on turmeric paper and moistened, gives an alkaline reaction.

Division 2. **Soluble in hydrochloric acid without leaving a perceptible residue; some also soluble in water; not gelatinized by evaporation.**

(Compare those of the former division which give only a weak alkaline reaction after fusion—*Kieserite, Kainite, Epsomite.*)

Durangite, fuses very easily; gives arsenic and fluorine reactions. Orange-red; streak yellowish.

Tschermigite, $(NH_4)_2AlS_4O_{16} + 24aq$; *Alunogen*, $AlS_3O_{12} + 18aq$; *Goslarite*, $ZnSO_4 + 7aq$. Fuse when first heated, and swell up to an infusible mass. All soluble in water; give sulphur reaction (par. 121). Heated on charcoal and treated with solution of cobalt, the former assume a blue, the latter a green, color (pars. 53, 54). The first, with caustic potash, gives the smell of ammonia; the second does not.

Chondrarsenite, $Mn_6As_2O_{11} + 3aq$, easily fusible, giving arsenic fumes on charcoal, and amethyst color to the borax bead. Color yellow.

Walpurgite, $Bi_{10}U_2As_4O_{24} + 12aq$; *Trögerite,* $U_2As_2O_{14}$ $+ 12aq$, give a green bead with salt of phosphorus, and the first gives the bismuth reaction with potassium iodide and sulphur.

Fauserite, $(Mn,Mg)SO_4 + 6aq$. Soluble in water; heated with phosphoric and nitric acids, gives a violet solution. Contains 40 per cent. of water.

Adamite, $Zn_4As_2O_9 + aq$, easily fusible, giving arsenic fumes on charcoal, with a coating of zinc. Color honey-yellow.

Struvite, $NH_4MgPO_4 + 12aq$, melts easily; yields water in matrass; with caustic potash, ammonia; and with hydrochloric acid, fumes of ammonium chloride. *Sussexite,* $(Mn,Mg)_2B_2O_5 + aq$. $H. = 3$. Fibrous; silky; gives a violet bead with borax.

Sassolite, H_3BO_3; **Boracite,** $Mg_7B_{16}Cl_2O_{30}$; *Hydroboracite,* $CaMgB_6O_{11} + 6aq$. Give the boric acid reaction (par. 75). Sassolite is soluble in alcohol; the others not; boracite yields no water, while the others do. Hydroboracite contains 26 per cent. water, and a similar mineral, *Szaibelyite,* $Mg_2B_4O_{11} + 3aq$, 7 per cent. *Lüneburgite,* $Mg_3P_2B_2O_{11} + 8aq$. Its nitric acid solution gives a yellow precipitate with ammonium molybdate.

Alabandite, MnS, and *Hauerite,* MnS_2, give strong manganese reaction (see par. 273).

Wagnerite, $MgF_2 + Mg_3P_2O_8$; **Apatite,** $3Ca_3P_2O_8 + Ca$ $(Cl,F)_2$. Moistened with sulphuric acid, impart a pale bluish-green color to the flame. Fusibility of wagnerite, 3–3.5 (with intumescence); of apatite, 5 (without intumescence); wagnerite is soluble in dilute sulphuric acid; apatite not. *Hebronite* is similar, and contains 4 per cent. of water.

Brushite, $HCaPO_4 + 2aq$, behaves in the wet way like

apatite, but yields 26 per cent. water. *Isoclasite*, $Ca_4P_2O_9$ + 5aq.

Amblygonite, $2AlP_2O_8 + 3(LiNa)F$. Fusibility 2; hardness 6. With difficulty soluble in concentrated sulphuric or hydrochloric acid. *Kjerulfine*, $2Mg_3P_2O_8 + CaF_2$.

Torbernite, $CuU_2P_2O_{12} + 8aq$; *Autunite*, $CaU_2P_2O_{12} +$ 10aq. Fuse readily, yield water, and give, with fluxes, the reactions of uranium sesquioxide (see Table II.). Soluble in nitric acid. The first gives a globule of copper with soda on coal. See Div. 3, p. 245.

Division 3. Soluble in hydrochloric acid, forming a stiff jelly, especially after partial evaporation.

Section 1. Before the blowpipe in a matrass give water.

Datolite, $H_2Ca_2B_2Si_2O_{10}$, yields but little water, and gives the boric acid reaction (par. 60).

Edingtonite, $BaAlSi_3O_{10} + 3aq$. The dilute hydrochloric acid solution gives a precipitate with sulphuric acid of barium sulphate. Sp. gr. 2.7.

Natrolite, $Na_2Al_2Si_3O_{10} + 2aq$. Fusibility 2; does not intumesce; hardness 5–5.5.

Scolecite, $CaAlSi_3O_{10} + 3aq$; **Laumontite,** Ca_2Al $Si_4O_{12} + 4aq$. Scolecite, on being heated, curls up like a worm, and finally melts to a bulky, shining slag, which in the inner flame becomes a vesicular, slightly-translucent bead; hardness 5.5; pyro-electric. Laumontite intumesces and fuses to a white translucent enamel; hardness 3.

Chalcomorphite, $CaO,Al_2O_3,SiO_2,H_2O,CO_2$, fuses in very fine splinters with difficulty, and curls up like scolecite.

Nearly related to scolecite, and showing a similar be-

22*   R

havior, are — *Mesolite*, (Ca,Na,AlSi,(

5; G.= 2.3; white, fibrous, and silky

2(Ca,Na,)AlSi,O, + 5 aq.   H.= 5; G.:

and vitreous; but they are not pyro-el

Phillipsite, (CaK,Na,AlSi,O,,, + 4 a

ity 3, with slight intumescence; occu

crystals; lustre vitreous; color white

dish.   *Gismondite*, (Ca,K,)Al,Si,O,, +

lated.   Trimetric, with forms often

octahedrons.   H. = 4.5.   Lustre spl

AlO,,CaO,Na,O,H,O,SO,SiO,.   Ash-g

ous lustre.   Fuses with intumescence,

the above in giving a precipitate in

acid solution with barium chloride.

Compare, in Div. 4, *Apophyllite, Okenite*, and

tinize with hydrochloric acid, 260.

Section 2. Before the blowpipe g
only traces.

(Compare *Datolite*, of the foregoin

Helvite, 3(Be,Mn,Fe),SiO, + (Mn,F

Mn,SiO,.   Distinguished from the othe

section by giving manganese reactions.

wax-yellow; hardness 6-6.5; of tephroi

ness 5.5–6.   *Danalite*, 3(Be,Mn,Fe,Zn

Zn)S, containing zinc, gives, with sod

slag of zinc, and with borax the iron

flesh-red to gray.

Compare *Willemite*, par. 303.

Hauynite, 2(Na,,Ca)AlSi,O, + (Na,,

pis-Lazuli, CaO,Na,O,AlO,,S,SiO,, a

color; give sulphur reaction (par. 12

the former, 4.5; of the latter, 3, forming a white glass.

Nosite, $2Na_2AlSi_2O_8 + Na_2SO_4$, and *Scolopsite*, AlO_3, $CaO, Na_2O, SO_3, Cl, SiO_2, H_2O$, of gray or brownish color; give sulphur reaction (par. 121). Fusibility of nosite, 4.5; of scolopsite, 3 (with intumescence like idocrase). The former crystallizes in dodecahedrons; the latter occurs granular-massive.

Sodalite, $3Na_2AlSi_2O_8 + 2NaCl$; *Eudialyte*, $6Na_2(Ca, Fe)_2(Si, Zr)_6O_{15} + NaCl$, give the chlorine reaction (par. 82). In the nitric acid solution, silver nitrate gives a precipitate. The former fuses to a transparent, colorless glass; the latter to a pistachio-green scoria or opaque glass. The dilute hydrochloric solution of eudialyte colors the turmeric paper orange-yellow; boiled with potassium sulphate and evaporated to crystallization, and then boiled with water, a precipitate of zirconia is formed, which makes the solution cloudy. Sodalite, H. = 5.5–6; G. = 2.3; eudialyte, H. = 5.5; G. 2.9.

Wollastonite, $CaSiO_3$, fuses quietly to a colorless, semi-transparent glass. The hydrochloric acid solution gives no, or only a very slight, precipitate with ammonium hydrate, but with the carbonate a bulky precipitate. See also *Pectolite*.

Nephelite, $(Na, K)_2AlSi_2O_8$.

Melilite, $(Na_2Ca, Mg)_{12}(AlFe)_2Si_9O_{36}$.

Meionite, $Ca_6Al_4Si_9O_{36}$.

With the solution of these minerals in hydrochloric acid, ammonia gives a precipitate. Meionite fuses, with intumescence, to a vesicular glass, which is not completely rounded by fusion. The others fuse quietly. The solution of melilite, after the separation of alumina with ammonium hydrate, gives a strong precipitate with am-

monium oxalate, while the solution of nephelite, trea___ed
in the same manner, gives no precipitate, or a very sli___ght
one.   Nephelite hexagonal; melilite dimetric.   *Elæol___ite,*
a variety of nephelite, has a greasy lustre.   Compare *Ca___n-
crinite.* The behavior of *Barsowite,* Ca,Al,Si_2O_8, is simil___la
to melilite, but fuses with more difficulty, and quietl_ ___ly
$F. = 4.$

Compare *Gehlenite,* Div. 5, p. 283, which is nearly infusible___ ___
Tachylite, Div. 5, p. 263; and *Willemite,* Div. 2, p. 276.

Division 4. Soluble in hydrochloric acid, with___ separation of silica, without forming a perfect jelly.

(It is sometimes necessary to treat the finely-pulverized mineral with___ ___
concentrated acid.)

Section 1. Before the blowpipe in a matrass give water.

Klipsteinite, MnO_3,MnO,SiO_2,H_2O, easily decomposed
by hydrochloric acid, evolving chlorine; and silica separates as a slimy powder.   Fuses to a black slag in the
oxidizing flame.   With phosphoric acid it gives a violet
solution.   By ignition gives 9 per cent. of water.

Apophyllite, $4(H_2CaSi_2O_6 + aq) + KF$;   **Pectolite,** $HNaCa_2Si_3O_9$;   *Okenite,* $H_2CaSi_2O_6 + aq.$ The
silica separates in the shape of gelatinous lumps.   After
the separation of the silica, the hydrochloric acid solution
gives no, or only a slight, precipitate with ammonia   Pectolite fuses to an enamel-like glass, with slight intumescence, and yields but little water; the others much.   Fusibility of apophyllite, 1.5, forming a white vesicular glass;
of okenite, 2.5–3; fuses with frothing, forming a porcelain-like mass.

Compare *Xonaltite* and *Sepiolite,* Div. 5, pp. 281, 282.

Analcite, $Na_2AlSi_2O_{12}$ + 2aq, gelatinizes like the preceding, in some varieties forming a perfect jelly. After the separation of the silica of the acid solution, ammonia produces a copious precipitate. Before the blowpipe, with the first action of the flame, it becomes opaque, but fuses quietly to a perfectly clear glass. The crystals are usually traperohedrons and cubes; not cleavable; yields 8 per cent. of water.

Pyrosclerite, $Mg_{22}Al_8Si_9O_{36}$ + 12aq; *Chonicrite*, $(CaMg)_{10}$ $Al_8Si_9O_{30}$ + 6aq; *Jollyte*, $(FeMg)_6Al_8Si_9O_{36}$ + 12aq, are distinguished from the other minerals of this section by their inferior hardness, 2.5–3. Chonicrite fuses from 3.5–4, with intumescence; has no cleavage; whitish; yields 9 per cent. of water. Fusibility of pyrosclerite, 4, without intumescence; cleavable in one direction; green; yields 11 per cent. of water. Jollyte fuses with difficulty; amorphous; brown; powder light-green.

Of similar constitution as pyrosclerite are—*Vermiculite*, $(Mg,Fe)_{22}Al_8Si_9O_{36}$ + 12aq, and *Jefferisite*, $Mg_4(Al,Fe)_8$ Si_8O_{20} + 6aq. They are both of brownish-yellow color, micaceous structure, pearly lustre, and before the blowpipe exfoliate remarkably; the former in worm-like forms.

Dudleyite, Kerrite, Maconite, Wilcoxite, and *Vaalite* belong with this section.

Brewsterite, $(Sr,Ba)AlSi_6O_{16}$ + 5aq, characterized by its hydrochloric acid solution giving a precipitate with sulphuric acid. It fuses with frothing and intumescence. Fusibility 3, and yields 13 per cent. of water.

Stilbite, $(Ca,Na_2)AlSi_6O_{16}$ + 6aq; *Hypostilbite*, $(Ca, Na_2)_3AlSi_6O_{20}$ + 12aq; Chabazite, $(H,K)_2CaAlSi_6O_{15}$ + 6aq; **Prehnite,** $H_2Ca_2Al_2Si_3O_{12}$. Fuse, with intumescence, to enamel-like masses. Prehnite yields but little water, losing by ignition only 4.3 per cent.; the others lose from ?? to ?? per cent. Chabazite is distinguished by its rhombohedral crystallization and imperfect cleavage. In stilbite and hypostilbite the cleavage is perfect in one direction. Stilbite is trimetric, and hypostilbite occurs in radiate-fibrous or columnar masses. *Mordenite*, $CaNa_2Al_2Si_5O_{12}$ – aq. H = 5: occurs in hemispherical reniform, or cylindrical concretions, with a fibrous structure, yields ?? per cent. water, and fuses without intumescence.

Mosandrite, $CaO.La_2O_3.DiO.CeO.Na_2O.TiO_2.SiO_2.H_2O$, and *Rosenbuschite*, $Na_2Ca_2.S.Zr_2O_4$ – 2aq. have hardness 4–4.5, and distinct cleavage. The first fuses with intumescence, then quietly to a yellowish-brown glass; fusibility 2.5–3: the second, 3: fuses quietly, giving a white porcelain bead. It is soluble in hydrochloric acid without gelatinizing, and gives the zirconia reaction, coloring turmeric paper orange-yellow. If boiled with potassium sulphate nearly to dryness, and water added, it gives a precipitate of zirconia. Mosandrite gives no precipitate; it contains ? per cent of water. Mosandrite, with salt of phosphorus in reducing flame, gives a violet color ...

Sepiolite [meerschaum], $Mg_2Si_3O_8$ – 2aq. See below. Deweylite, $Mg_2Si_2O_?$ – ?aq. Distinguished by being much less fusible than the preceding: fusibility 5. The former absorbs water with great avidity: the latter not. Sepiolite contains 10 per cent. of water, and deweylite 20 per cent.

Cordierite, $Al_2O_3.FeO.MgO.SiO_2.H_2O$. Amorphous; fusibility 2.5, forming a thick, black, brilliant glass: color brownish-black. Decomposed by hydrochloric acid with difficulty. The solution gives a greenish-gray precipitate with ammonia. It contains 4 per cent. of water.

Section 2. Before the blowpipe in a matrass give no water, or only traces.

(Compare *Pectolite*, *Chonicrite*, and *Prehnite*, of the preceding section.)

Many specimens of lapis-lazuli do not form a complete jelly, but may be recognized by their blue color.

Cryophyllite, $(K,Li)_{12}Fe_3(Al,Fe)_4Si_{21}O_{67}$. Micaceous; fuses easily in the flame of a candle, giving the flame a lithia reaction.

Tachylyte, $Na_2O,CaO,MgO,FeO,AlO_3,SiO_2H_2O$, fuses easily (2.5) and quietly to a black shining glass. Hardness 6.5; color black. Decomposed with hydrochloric acid, the silica separates in lumps. The solution, boiled with tin, does not become violet, which is the case with the two following minerals:

Schorlomite, $Ca_6Fe_2(Si,Ti)_{12}O_{30}$, and *Tscheffkinite*, CeO, FeO,CaO,TiO_2SiO_2. Fusibility 3–4. The first fuses quietly, the second with much effervescence, to a black glass or a grayish mass. The first is decomposed with difficulty by hydrochloric acid, and the silica separates as a slimy powder; the second is easily decomposed, and the silica separates in gelatinous lumps. Color of both black; powder gray.

Ivaarite is similar to schorlomite.

Wernerite, $(Ca,Na_2K_2)Al_2Si_3O_8$, and *Porcellanite*, SiO_2 Al_2O_3,CaO,Na_2O,Cl_n, fuse easily at 2.5, with intumescence, to a white vesicular glass, which is not easily rounded. They are quite cleavable in two directions. With wernerite belong *Nuttalite*, *Glaucolite*, and *Stroganovite*.

Wöhlerite, $CaO,Na_2O,SiO_2ZrO_2Cb_2O_5$, fuses easily at 3 to a light-green, very blebby glass. Decomposed with hydrochloric acid, with separation of silica in flocks; the solution, strongly boiled with tin, becomes a beautiful blue (*columbium*), and on the addition of water, a blue filtrate. This solution colors turmeric paper orange-yellow. The mineral is wine-yellow, honey-yellow, or brownish-red. *Eucolite* probably belongs here.

Labradorite, $(Ca,Na_2)AlSi_3O_{...}$ and **Anorthite,** Ca, $AlSi_2O_8$, fuse quietly, without intumescence, forming a thick colorless glass. Hardness of the former, **6**; of the latter, **6–7**. Anorthite fuses with more difficulty (4.5) than labradorite (3.5). Labradorite cleaves in two directions with an angle of 94°, and on the perfect cleavage planes shows striæ, on the others none, and frequently a play of colors, blue and green, also red and yellow. Anorthite has perfect cleavage at an angle of 94° 12'. The labradorite is not wholly decomposed with hydrochloric acid.

Grossularite [some varieties]. Fusibility **3**; not cleavable.

Titanite [sphene] (Div. 6, p. 269), some varieties, see below. Gives titanium reactions (par. 125). See *Danburite*, Div. 6, p. 265, which gives a fine green flame; also *Tephroite*, Div. 3, p. 258. which gives an amethyst color to the borax bead.

Microsomnite, $(K_2Ca)Al_2Si_4O_8,CaSO_4,NaCl$. Colorless; vitreous; $H. = 6$; $Fus. = 5$; gives the chlorine reaction with copper oxide and salt of phosphorus.

Division 5. **Little affected by hydrochloric acid.** **Before the blowpipe give an amethyst color to the borax bead** (*manganese*).

Carpholite, $H_4Mn(AlFeMn)Si_2O_{10}$, occurs only in fibrous, radiated tufts. Color straw-yellow; silky; yields water, 11 per cent.; fusibility 2.5–3.

Ardennite, of similar composition, with 9 per cent. of

VO$_3$; fibrous; stellated; of brownish-yellow color; yields water, 5 per cent.; fusibility 2.

Spessartite [Manganese Garnet], (Mn,Fe)$_3$AlSi$_3$O$_{12}$. Color brownish-red; fuses without intumescence at 3; not cleavable.

Piedmontite, H$_2$Ca$_4$(MnFeAl)$_3$Si$_6$O$_{25}$. Fusibility 2–2.5; intumesces; cleavage quite distinct in one direction; less so in a second; color cherry-red to reddish-black.

Rhodonite, MnSiO$_3$. Fusibility 3, without intumescence, when pure; color rose-red; cleavable (par. 275). It cleaves at an angle of 92° 55′. *Richterite* [manganese amphibole], cleaves at 124°.

Compare *Axinite*, p. 267.

Division 6. Not belonging to either of the preceding divisions. All are silicates except *Scheelite*, and are not decomposed, or only partially, by hydrochloric acid.

Danburite, CaB$_2$Si$_2$O$_8$. Fusibility 3, and gives a fine green color to the flame (*boric acid*). The bead is clear while hot and cloudy when cold. Yields no water in the matrass. *Howlite*, Ca$_4$B$_{10}$Si$_2$O$_{23}$ + 5 aq, is closely related, but yields water in the matrass.

Scheelite, CaWO$_4$. Fusibility 5. Soluble in hydrochloric acid, leaving a greenish-yellow or lemon-yellow residue of tungstic acid, which is soluble in ammonia, and which gives, with salt of phosphorus, the characteristic reaction of tungstic acid (see Table II.). If the residue of tungstic acid is boiled with phosphoric acid till it begins to fume, after cooling a blue mass is formed, which gives a colorless solution with water. If iron filings be added to this and shaken for some time, it becomes intensely blue. G. = 6.

23

Lepidolite, $(KLi)_6Al_4Si_{12}O_{30}$, and *Cookeite,* $(Li_2O,K_2O, Al_2O_3,SiO_2,H_2O)$, are micaceous, splitting very easily in one direction. Fusibility of lepidolite is 3, colors the flame crimson, and gives little or no water in the matrass. Cookeite intumesces, colors the flame crimson, but yields much water in the matrass.

Gümbelite, SiO_2,Al_2O_3,K_2O,H_2O, occurs in short fibres. Before the blowpipe swells into a fan-shaped mass, fuses in thin fibres, and gives in the matrass 7 per cent. of water. Not attacked by hydrochloric or sulphuric acid.

Thermophyllite, $Mg_3Si_2O_7 + 2$ aq, *Euphyllite,* $(AlNa_4 K_6)_4Si_9O_{36} + 4$ aq, and **Margarite,** $H_2CaAl_4Si_2O_{12}$, are all micaceous in structure. The first intumesces before the flame and yields much water; the others fuse without intumescence (4-4.5) and yield little water. Their laminæ are not elastic. Euphyllite is easily decomposed by sulphuric acid, and margarite with difficulty.

Compare *Muscovite* and *Biotite.*

Petalite, $(Li_6Al_9)Si_6O_{25}$, and **Spodumene,** $(Li_6Al_4)Si_9O_{9}$, do not possess as perfect a cleavage as the preceding, and greater hardness (6.5). Specific gravity of petalite 2.4-3; of spodumene, 3.1. Both give the lithia reaction (par. 103). Spodumene fuses, with intumescence, to a clear or white glassy globule; petalite fuses quietly to a white enamel. *Castorite* belongs here, which alone gives a distinct red color to the flame.

Leucophanite, $4NaF + 3(CaBe)_4Si_3O_{10}$, fuses easily and quietly to a transparent colorless glass. Cleavage very marked in one direction. H.$=3.5-4$. If heated, phosphoresces with a reddish-violet light; also if struck with a hammer in the dark.

Wilsonite, $Al_2O_3,K_2O,MgO,SiO_2,H_2O$. Fusibility 2,

swelling up to a whitish glass; yields water in a matrass; H.$=$3; cleaves at right angles.

Nohlite. Nb_2O_5,U_2O_3,YO,H_2O, compact, brownish-black, fuses with difficulty, and gives in a matrass 4.5 per cent. of water; otherwise in chemical properties similar to samarskite.

Sordawalite. Fusibility 2.5; amorphous; brownish-black. See Div. 4, p. 262.

Diallage, $Ca,Mg,etc.,SiO_2$. Fusibility 3.5; characterized by its pearly metallic lustre; cleaves easily in one direction.

Harmotome, $Ba,Al_2Si_5O_{14} + 5$ aq, distinguished from most of the other minerals of this division by yielding water in a matrass. In the partial solution in hydrochloric acid, sulphuric acid gives a precipitate with the barium. Occurs usually in twin crystals.

Axinite, $(Ca,Fe,K_2),(AlFeB)_3Si_8O_{32}$; **Tourmaline,** $AlO_3,BO_3,FeO,MgO,MnO,K_2O,Na_2O,Li_2O,SiO_2,F$, give the reaction of boric acid (par. 61). Axinite fuses readily, with intumescence, to a dark-green glass. Different varieties of tourmaline show different blowpipe characteristics, but all are pyro-electric. Hardness of axinite and of tourmaline 6.5–7.5. Heat does not develop electricity in axinite.

Pyroxene. General formula, $RSiO_3$. R may be Ca, Mg,Fe,Zn,Mn,K$_2$,Na$_2$(Al,Fe,Mn). Two or more of these bases are usually present. Calcium is always present, and constitutes a large per cent. of the mineral. H.$=$5–6; G.$=$3.2–3.5. Monoclinic. Cleavable nearly at right angles of 93° and 87°. Occurs in thick, stout prisms; massive, granular, fibrous, and lamellar; color shades of green, white to brown or black, through *bluish* shades, but not *yellow;* lustre vitreous and somewhat pearly.

II. Dark-colored.

Augite, (Ca,Mg,Fe,Al). G.= 3.3–3.5.
black and black crystals; occurring in erup

Hedenbergite, (Fe and Ca), color black; cry
lar-massive. *Jeffersonite*, (Ca,Fe,Zn,Mg), col
black; crystals often very large.

Polylite and *Hudsonite* belong here. The
ored varieties are more fusible than the light-c
the globule obtained is colored black by the i

III. *Diallage* is a thin, foliated variety, often
serpentine and other rocks. Differs from bron
persthene in crystalline form, and in being fu

Amphibole. Formula as for pyroxene, an
characteristics the same. H.= 5–6; G.= 2.9–
clinic. Cleavable at 124½° and 55° 30', by
distinguished from pyroxene. Often in lon
flat, rhombic, also six-sided, prisms; also
coarse and fine fibrous, lamellar, and gran
white to black, passing through various shade
lustre vitreous, with cleavage face sometimes

Varieties:

I. Light-colored. G. 2.9–3.3.

Tremolite and *Grammatite*, (Mg,Ca). Color white to dark-gray; in crystals, columnar, fibrous, compact, granular-massive.

Actinolite, (Mg,Ca,Fe). Light-green varieties; in bright-green crystals, columnar, fibrous, often radiated, and also granular-massive.

Asbestus, slender, flax-like fibres, green, gray, or white. *Amianthus*, white, silky. Ligniform asbestus, mountain leather, and mountain cork, usually white or grayish-white, belong here.

Nephrite is a tough, compact variety, closely related to tremolite. These light-colored varieties contain little or no alumina or iron.

II. Dark-colored. G.=3–3.4.

Hornblende, (Mg,Ca,Al,Fe), black and greenish-black crystals and massive. *Pargasite*, dark-green, short and stout crystals. *Cummingtonite*, color gray or brown; usually fibrous, and often radiated.

III. *Smaragdite*, a thin, foliated variety of a light-green color, resembling common diallage.

Titanite, $CaTiSiO_5$. Fusibility 3. H.=5–5.5. Monoclinic; gives the titanium reaction (par. 125); imperfectly soluble in hydrochloric acid.

Guarinite, $CaTiSiO_5$, of similar composition, but dimetric.

Keilhauite, containing 28 per cent. of TiO_2; also alumina and yttria. Fuses, with intumescence, to a black shining glass. Yields with borax an iron-colored glass, which, in the inner flame, becomes blood-red. Reaction of manganese with soda. Decomposed by hydrochloric acid.

23*

Orthoclase, $K_2AlSi_6O_{16}$; **Albite,** with potash by soda, $Na_2AlSi_6O_{16}$. Hardness 6; fuse with mescence. Fusibility of orthoclase 5; of albit latter colors the flame yellow. Not soluble With cobalt solution become blue on the edges (Orthoclase cleaves in two directions at right ang albite at 93° 30', and show striæ on one surface.

Oligoclase, $(Ca, Na_2, K_2)AlSi_3O_{16}$, is more fus albite (3.5). It sometimes resembles labrador unlike it, is not materially acted upon by aci cleavage surface shows the striæ in a marked d the mineral is powdered, mixed with ammoni ride, and ignited in a platinum dish, then boi hydrochloric acid, neutralized with ammonia, a ed, the lime in the filtrate may be precipitated I moninum oxalate. *Tschermakite,* $(Na_2Ca)Al_2Si$

Hyalophane, $(Ba, K_2)AlSi_2O_{12}$, is very similar minerals, but, if fused with potash, treated wit chloric acid and water, the solution gives a p of barium sulphate with sulphuric acid.

Zoisite, $H_2Ca_4(AlFe)_6Si_6O_{26}$, and **Epidote,** 1 $F)_3Si_6O_{26}$. Hardness 6.5; fusibility 3–3.5; fuse tumescence—zoisite to a white or yellowish slag; to a black or dark-brown slag. After fusion they g with acid. Color of zoisite gray, yellowish-gray, white; of epidote, green.

Garnet [var. *Grossularite*], $Ca_3AlSi_3O_{12}$; [var. $(Mg, Ca, Fe, Mn)_3Al_2Si_3O_{12}$; and **Vesuvianite,** Ca Fe, or $H_2, K_4Na_2)Al$, or $FeSi_7O_{26}$. Hardness 6.5–7 bility of lime-garnet and vesuvianite, 3; of pyr Vesuvianite possesses cleavage; the others not. gives, with the fluxes, the chromium reactions

others green, yellowish-brown, hyacinth-red, and white. The *Edelforsite* and *Sphenoclase* of Von Kobell belong here.

Monzonite resembles grossularite, but does not gelatinize after fusion, and is not decomposed by hydrochloric or sulphuric acid. Color gray-green. See also *Emerald, Euclase, Iolite, Biotite,* and *Muscovite.*

Obsidian, Pitchstone, Pearlstone, and Pumice, $SiO_2, AlO_3, FeO_3, CaO, MgO, Na_2O, K_2O, H_2O$, are amorphous. Fusibility 3.5–4; fuse, with intumescence, to porcelain-like masses or white vesicular glasses. Lustre of obsidian glassy; of pitchstone, greasy; of pearlstone, pearly. Pumice is characterized by its porosity. Pitchstone usually yields water in the matrass.

CLASS III. Infusible, or fusibility above 5.

Division 1. After ignition, moistened with cobalt solution and again ignited, assume a bright-blue color (*alumina*).

(Some minerals should be first calcined and pulverized.)

With the hard, anhydrous minerals of this division the color is best seen by reducing the substance to a fine powder and moistening this with the cobalt solution. The color appears only after cooling, and by daylight.

Section 1. Give much water in a matrass.

Ralstonite, Al, F, etc., gives off hydrofluoric acid when warmed with sulphuric acid.

Alunite, $K_2Al_3S_4O_{22} + 6aq$, and **Aluminite,** $AlSO_6 + 9aq$, with soda on coal give a sulphur reaction, which is not the case with the following minerals. Aluminite is readily soluble in hydrochloric acid; alunite, not visibly affected. By calcination alunite loses 13 per cent. water,

aluminite 47, and a similar mineral, *Felsobær*
$O_rSO_s + 10$ aq).

Pissophanite, $AlO_r, FeOSO_s, H_sO$, blackens
to which it gives a greenish tinge, burns and fal
Aluminite is white and opaque, while the latter
and transparent.

(See also *Kalinite*, *Alunogen*, and *Tscherm*
are soluble in water, while the foregoing are 1

Plumbogummite, see par. 264; **Calamine**, se

Gibbsite, H_6AlO_4; **Diaspore**, H_sAlO_4;
$(Mg_sCa_sAl_sFe_s)SiO_s$; and *Pholerite*, Al_sSi_sO
Gibbsite is easily soluble in potassium hydrat
by ignition 34.5 per cent. water. The other
uble in potash. Distinct cleavage in one direc
bertite loses 4½ per cent. water by ignition.
yellow. Diaspore and pholerite lose 15 per
may be distinguished from the other miner
hardness—diaspore, 6; pholerite, 1. The la
curs in scales, with a mother-of-pearl lustre.

The following minerals of similar composi
the most part soluble in caustic potash. If a
nitric acid is added to the solution and boile
monium molybdate, a yellow precipitate is f
the closed tube give the phosphoric acid re
par. 110).

Wavellite, $Al_sP_4O_{so} + 12$ aq, loses by ign
cent. of water.

Evansite, $Al_sP_sO_{14} + 18$ aq, loses by ignit
cent. of water.

Peganite, $Al_rP_sO_{11} + 6$ aq, loses by ignition
of water.

Fischerite, $Al_sP_sO_{11} + 8$ aq, loses by ignit

Berlinite, $2AlP_2O_8 + aq$, loses by ignition 4 per cent. of water.

Zepharovichite, $AlP_2O_8 + 6aq$, loses by ignition 27 per cent. of water.

Trolleite, $Al_4P_6O_{27} + 3aq$, loses by ignition 6 per cent. of water.

Sphærite, $Al_5P_4O_{25} + 16aq$, loses by ignition 24 per cent. of water.

Redondite, $AlO_3, FeO_3, P_2O_5, H_2O$, loses by ignition 23 per cent. of water.

Tavistockite, $Ca_3AlP_2O_{11} + 3aq$, loses by ignition 12 per cent. of water.

Amphithalite, AlO_3, CaO, P_2O_5, H_2O, loses by ignition 12 per cent. of water.

Cæruleolactite, $Al_3P_4O_{19} + 10aq$, loses by ignition 21 per cent. of water.

These minerals are of various grayish shades of green, yellow, red, and brown to white, and vitreous or pearly lustre. H.$= 3$-6; G.$= 2$-3.

Allophane, $AlSiO_5 + 5aq$; *Halloysite,* $AlSi_2O + 4aq$; *Samoite,* $Al_2Si_3O_{12} + 10aq$; *Collyrite,* $Al_2SiO_8 + 9aq$. Decomposed by hydrochloric acid, with separation of gelatinous silica. Hardness of allophane, 3; gelatinizes completely; often colors the flame green, showing the presence of copper; and loses by ignition 42 per cent. of water. Amorphous. The hardness of samoite is 4; structure laminated, and loses by ignition 30 per cent. The hardness of the others is 1-2. Halloysite loses on ignition 16 per cent. of water; collyrite, 33.5.

Cimolite, $Al_2Si_9O_{24} + 6aq$; **Kaolinite,** $AlSi_2O_7 + 2aq$, are very soft and earthy, and but little affected by acids; lose on ignition from 12 to 16 per cent. of water. Nearly related to these minerals are the various varieties of com-

mon *clay*, some varieties of *Lithoma*
of water), and *Schrötterite*, with 35 |
loschite and *Halloysite*, with ~~aq. 46~~ |
clays become plastic with water; th
fall to pieces.

Compare also *Lazulite, Svanbergite,*
Wörthite, Myelin, Agalmatolite, which yi
only a very little. Compare also *Ripido*

Section 2. Before the blowpipe no water, or but a trace.

Alumian, AlS_2O_9.   Before the b
gives the sulphur reaction.

Lazulite, $(Mg,Fe)AlP_2O_9 + H_2O$.
of phosphoric acid (par. 110).   B
color and becomes white.   Not af

Svanbergite, P_2O_5,SO_3,AlO_3,CaO
gives the sulphur reaction; colo
brown.

Willemite, Zn_2SiO_4.   With cob
becomes blue, and green in spot
hydrochloric acid (see par. 303).

Myelin, Al_2SiO_5; *Agalmatolite*, Si
rophyllite, $AlSi_3O_9 + H_2O$.   Are
1–2.   Pyrophyllite is foliated like t
pipe swells up and spreads out int
creasing to about 20 times its forr
pact varieties do not exfoliate.   The
before the blowpipe.   Myelin is pai
hydrochloric acid; agalmatolite not

Westonite, AlO_3,SiO_2,H_2O.   Colo
like pyrophyllite, but is dull, and ne

Muscovite, $H_2AlSi_2O_6$.   Cleavag

rection; folia elastic; does not swell perceptibly before
the blowpipe; fusible in very thin laminæ; with cobalt
solution is blue only in spots; not affected by acids;
hardness 2.5.

Brandisite (variety of **Seybertite**). Cleavable in one
direction. H. 4–5. Before the blowpipe fresh pieces
become grayish-white and cloudy, and then, moistened
with cobalt solution and ignited, become distinctly blue.
Decomposed by concentrated sulphuric acid.

Andalusite, $AlSiO_5$; **Cyanite,** $AlSiO_5$, are but little
affected by acids. Cyanite occurs generally in bladed
crystals; cleavage very perfect at 106°. H.$=5$–7.
Color blue, white, gray, black. Hardness of andalusite
7.5. Andalusite with salt of phosphorus is decomposed,
leaving a skeleton of silica in the bead. It cleaves in two
directions at 91½°. H.$=7.5$; G.$=3.5$.

Sillimanite, Wörthite, Monrolite (vars. of **Fibrolite**),
$AlSiO_5$, are closely related.

Topaz, $AlSi(O,F_2)_5$; **Rubellite** [Tourmaline], (Li,
Na,K)$_6$Al$_6$B$_2$Si$_9$O$_{45}$. Not affected by acids. Not completely soluble in salt of phosphorus; the glass becomes
opalescent on cooling. Topaz on being ignited remains
transparent and does not swell. Fused in the open tube
with salt of phosphorus, gives the fluorine reaction.
Topaz is cleavable in one direction. H.$=8$. Rubellite
on being ignited becomes white and swells; fused with
acid potassium sulphate and fluorite, gives a green flame
(boric acid). Is pyro-electric. No cleavage. H.$=$
6.5; G.$=3$.

Corundum [Sapphire], AlO_3; **Chrysoberyl,** $BeAl$
O_4. Not affected by acids. When pulverized, slowly
but completely soluble in salt of phosphorus; the glass
does not opalesce on cooling. Hardness of chrysoberyl

8.5; G.$=$3.7; of corundum, the former usually green; of the brown.

Compare *Spinel*.

(Some varieties of *Leucite* as cobalt solution, but its hardness is also, in fine powder, takes a blue with cyanide of potassium globu takes a pale-blue color with a red

Division 2.   Moistened with
ignited, assume a green cc

It is sufficient to heat to red this division give a coating of par. 34.

Smithsonite, see par. 302.

Hydrozincite [Zinc Bloom] solves readily in hydrochloric a the solution gives with ammonia uble in an excess of the reage matrass.

Willemite, see par. 303; C Gelatinize with hydrochloric a water, willemite not. With cc green color only in spots.

(See *Sphalerite* and *Goslarite*,

Division 3.   After ignition
action, and change to red of moistened turmeric par

Brucite, H_3MgO_2; *Hydrodo* H_2O; **Hydromagnesite,** Mg_4C_3 water in a matrass, unlike the

division. Brucite dissolves in hydrochloric acid without effervescence, hydromagnesite with effervescence. The concentrated hydrochloric acid solution of the hydromagnesite is not precipitated by sulphuric acid, while the latter yields a heavy precipitate. *Predassite*, $2Ca$ $Co_3 + H_2MgO_2$, and *Pencatite*, $CaCo_3 + H_2MgO_2$, are similar. in behavior to the hydrodolomite. *Pyrochroite*, $H_2(MnMg)O_2$, is similar in reactions to brucite, but boiled with concentrated phosphoric acid, gives, on addition of nitric acid, a violet-red solution. *Lancasterite* is a mixture of brucite and hydromagnesite. *Nemalite* is a fibrous variety of brucite, of silky lustre.

Calcite, $CaCo_3$; **Aragonite,** $CaCo_3$. Dissolve readily and with effervescence in dilute cold hydrochloric acid; the concentrated (but not the dilute) solution gives a precipitate with sulphuric acid. Aragonite falls to powder before the blowpipe, calcite not. Calcite, H.$=3$; G.$=2.6$–2.8; aragonite, H.$=3.5$–4; G.$=2.9$–3. (See *Strontianite.*)

Dolomite, $(Ca,Mg)Co_3$; **Magnesite,** $MgCo_3$. Do not, or but slightly, effervesce with cold dilute hydrochloric acid, but dissolve readily on application of heat. The concentrated solution of the former gives a precipitate with sulphuric acid, that of the latter not.

Compare *Siderite*, see par. 254, and *Rhodochrosite*, see par. 274.

Strontianite, $SrCO_3$; **Barytocalcite,** $(Ba,Ca)CO_3$. Dissolve with effervescence in dilute hydrochloric acid; the solution, even if largely diluted with water, gives a precipitate with sulphuric acid. Strontianite colors the flame red, par. 59; barytocalcite, yellowish-green, par. 60.

Compare *Yttrocerite;* also *Talc* and *Muscovite*, which, after ignition, sometimes give an alkaline reaction.

24

Division 4. Completely solu
hydrochloric or nitric aci
ising by evaporation or lea
residue of silica.

Lithiophorite,   MnO,CuO,CoO
H₂O.   Color bluish-black.   Col
red.

Ludwigite, FeO₃,FeO,MgO,Ba(
cultly fusible.   With sulphuric ac
green flame of boric acid.

Cervantite, Sb₂O₃ + Sb₂O₅.   B
coal, infusible, but with soda eas
antimony.   Color yellowish.

Stibiconite, SbO₂ + H₂O, and
are similar.   They yield water in
5 per cent., and the latter 15 per

Siderite, see par. 254; **Rhodo**
Zaratite, see par. 284.   Dissol\
acid with effervescence.

Mesitite, Mg₂FeC₃O₉.   Black\
netic before the blowpipe.   Slig\
cold by acids; but, if powdered\
effervescence in hot hydrochloric
white to brown.   Streak, nearly v

Ankerite, (Ca,Fe,Mg)CO₃, is \
dissolved in aqua regia, the iro
monia, a heavy precipitate will
of ammonium oxalate.

Hydrotalcite, AlO₃,MgO,H₂O,(
matrass.   Does not become mag
flame.   In powder, effervesces \
and dissolves completely.   If the

with sodium carbonate and filtered, ammonium oxalate gives no precipitate in the filtrate, but a precipitate is formed by hydro-disodium phosphate and ammonia.

Parisite, $3(Ce,La,Di)CO_3 + (Ca,Ce)F_2$, is slowly soluble in hydrochloric acid with effervescence. The solution, not too acid, gives a white precipitate with oxalic acid, which becomes brick-red by ignition (*cerium oxide*).

Limonite, see par. 248; **Göthite,** see par. 249.

Turgite, $H_2Fe_2O_7$, forms a brownish-red powder, and loses by ignition 5.7 per cent. of water (see par. 250).

(See also *Hematite*, which in some varieties is without metallic lustre; readily distinguished by its red streak.)

Sphalerite, see par. 300; *Greenockite*, CdS. Dissolve in hydrochloric acid with evolution of sulphuretted hydrogen. Give the sulphur reaction, par. 121. Greenockite gives on charcoal a coating of cadmium oxide, par. 35, the others of zinc oxide, par. 34. *Marmatite*, FeS + 3 ZnS, gives, after calcination with the fluxes, the reactions of iron.

Wad, see par. 273; **Zincite,** see par. 301.

Asbolite (var. of *Wad*), see par. 273. Some varieties are fusible.

Uraninite, U_3O_8; *Zippeite*, $U_3S_2O_{15} + 12H_2O$. Give with the fluxes the reactions of uranium sesquioxide (Table II.). Give with nitric acid a yellow solution, in which ammonia produces a sulphur-yellow precipitate. Uraninite is black; zippeite, yellow. G. of uraninite, **6.5.**

Turquois, $Al_2P_2O_{11} + 5$ aq with Cu. Color sky-blue to green. Gives the copper reaction, par. 91. Mostly soluble in caustic potash, leaving a residue of brownish color containing copper. The solution in nitric acid

Apatite, 3 CaP₂O₈... (partly obscured)

acid reaction; par... ...

acid.   If the solution

lead phosphate is for

cium oxalate with am

Monazite,   5(Ca, La

Gives the phosphoric

in hydrochloric acid.

brown color.   G. = 5.

Childrenite, (**FeMn**)

phoric acid reaction, p

reaction of iron and m

in hydrochloric acid.

ened with sulphuric aci

much water.   Sp. gr. 3

Polycrase, YO, TiO₂,

itates, but infusible.

verized mineral with c

mass in water, neutrali

acid, a precipitate is

excess of concentrate

gives a cloudy blue sol

after the addition of a

turmeric paper orange-

Fluocerite, Ce₃F₉, a

4 aq, give the reactio

cerium sesquioxide (T

carbon dioxide when t

Division 5. With hydrochloric acid gelatinize, or decompose with separation of silica.

Section 1. Before the blowpipe give water in a matrass.

Dioptase, H_2CuSiO_4 emerald green; **Chrysocolla**, see par. 234, blue to green, $H. = 2-4$; *Cyanochalcite*, CuO, P_2O_5, SiO_2, H_2O, azure-blue, $H. = 4.5$, lustre dull. Behave alike before the blowpipe; the first gelatinizes with acids, the latter not.

Uranotile, $Ca_2U_6Si_5O_{30} + 15$ aq. Color lemon-yellow, lustre vitreous. Acicular crystals.

Xonaltite, $4CaSiO_3 + H_2O$. Massive; very hard; white to gray; yields water; infusible (?); decomposed by hydrochloric acid, in which solution of ammonium oxalate gives a heavy precipitate, but ammonia none.

Thorite, $Th_2SiO_4 + H_2O$; *Cerite*, $(Ce,La,Di)_2SiO_4 + H_2O$. Gelatinize with hydrochloric acid. The solution of cerite, not too acid, gives, with ammonium oxalate, a white precipitate, which becomes brick-red if ignited on platinum (cerium oxide). Color of thorite is black; streak dark-brown; hardness 4.5–5; of cerite, brown to red, passing into gray; streak white; hardness 5.5. Their sp. gr. is 4.9–5.4.

Chloropal, $Fe_2Si_3O_9 + 5$ aq; *Wolchonskoite*, $MgO, Al_2O_3, Cr_2O_3, Fe_2O_3, SiO_2, H_2O$; and *Genthite*, $H_4(Ni,Mg)_4Si_3O_{12}$. Amorphous, with resinous lustre (see par. 285). Wolchonskoite is dark sea-green, and gives with borax an emerald-green bead, which continues when cold. The others are yellowish-green. Chloropal gives a green bead, which fades on cooling, and genthite a violet bead in oxidizing flame, becoming gray in reducing flame. If the mineral is powdered and moistened with potassa, the chlo-

24 *

ropal becomes black ~~without~~ [illegible]
after boiling until concentrated; and a
changed. The hydrochloric acid solu
ammonium in excess becomes azure-bl
off water in the closed tube and black

Compare *Gillingite*, Div. 5, p. 250,
5. Sect. 3. Become magnetic by ig
composed by hydrochloric acid. G
amorphous; xylotile is light- or dark-
woody structure.

Sepiolite, $Mg_3Si_2O_8 + 2aq$, gelat
chloric acid; very light; G.$= 1.5$; a
great avidity; gives the magnesia re
solution (par. 53); before the blowpi
shrinks; forms a jelly-like mass with

Bastite or *Schiller-spar* [impure Se
tile, $Mg_3Si_2O + 2aq$, possess a meta
the former is massive, cleavable; the
ignition schiller-spar becomes brown;
Both are decomposed by hydrochlo
readily by sulphuric acid, without gel
ite is greenish-white, similar to chrys

Cerolite, $H_4Mg_3Si_2O_7 + H_2O$. An
3; G.$= 2.3$. Color greenish, yellowis
the blowpipe blackens, but does not I
cobalt solution, a pale flesh-red color.
30 per cent.

Serpentine, $Mg_3Si_2O + 2aq$; ofter
Cr. Decomposed by concentrated hydr
out gelatinizing. Usually massive and
3-4; loss by ignition, 12 to 13 per cen
position, and showing a similar behav
ing minerals, which, however, possess c

and cleavage: *Picrophyll*, fibrous; greenish-gray; hardness 2.5; loss by ignition, 10½ per cent. *Picrosmine*, greenish-white, dark-green, gray; hardness 2.7; loss by ignition, 9 per cent. *Marmolite*, greenish and bluish-white; hardness 2.5–3; loss by ignition, 15.7 per cent. *Antigorite*, hardness 2–5; loses by ignition, 4 to 6 per cent. *Penninite*, $Mg_2Al_6Si_3O_{14} + 4aq.$ H.= 2.5; lustre pearly; color green, gray, red; crystals often tabular and in regular groups; gives little water and exfoliates.

(See also *Chlorite* and *Ripidolite*, which are with difficulty decomposed by concentrated hydrochloric acid.)

Monradite, $4(Mg,Fe)SiO + H_2O$; *Neolite*, $MgO,AlO_3,$ SiO_2,H_2O. Decomposable by concentrated hydrochloric acid without gelatinizing; loss by ignition, 4 to 6 per cent. Monradite, hardness 6; neolite, in silky fibres, or massive; hardness 1.

(See also some varieties of *Seybertite*, $Mg_6Ca_6Al_2Fe_2SiO_6$. Hardness 4–5; lustre pearly, submetallic; color yellow, reddish-brown, copper-red.)

Section 2. Before the blowpipe in a matrass give no water, or but a trace.

Gadolinite, $SiO_5(Y,Fe,Ce,Be)_3$; *Gehlinite*, $Ca_3(AlFe)Si_2$ O_{10}, gelatinize with hydrochloric acid. Gadolinite swells before the blowpipe into cauliflower-like masses, and sometimes exhibits a vivid glow; thin splinters fusible on the edges; color black to blackish-green; hardness 6.5–7; G.= 4–4.3. Gehlinite is also fusible in very thin splinters; color gray to grayish-white; hardness 5.5–6; G. =3.

Chrysolite, $(Mg,Fe)_2SiO_4$; H.= 7; **Chondrodite**, Mg_3 Si_3O_{14}, with part of the oxygen replaced by fluorine. H. =6.5; gelatinize with hydrochloric acid; color of the

angle of divergence does not exceed 5°, and seldom 1°. Turned in the stauroscope, the black cross is not changed, while with the others it is changed with various colors. The optic axial angle of muscovite is 44°–78°, of margarodite the same, and of phlogopite 3°–20°, seldom less than 5°. The laminæ of biotite and muscovite are elastic, of talc not. Soapstone, or *Steatite*, is a massive, usually compact, variety of talc; very greasy to the feel, or like soap. (See also *Pyrophyllite*.) Margarodite, $K_2Al_4Si_6$ $O_8 + H_2O$, and *Phlogopite*, $K_2Mg_6Al_2Si_5O_{20}$, are decomposed by sulphuric acid. Margarite, $Ca_2Al_4Si_2O_{11} +$ H_2O, with pearly lustre, and *Œllacherite* $K_2(Ba,Mg)Al_2$ Si_3O_{12}, with 5½ per cent. baryta, are nearly related to the muscovite.

Prochlorite, $H_{10}(FeMg)_{10}Al_3Si_6O_{42}$; *Delessite*, Mg,Fe AlO_3,FeO_3SiO_3,H_2O; Ripidolite, $Mg_5Al_2Si_3O_{14} + 4aq$. Lose by ignition 12 per cent. of water. Cleavage eminent in one direction; laminæ not elastic (both often massive-granular). Hardness of prochlorite, 2; of ripidolite, 2–2.5. Delessite has a short fibrous structure. $H. = 1–2.5$. Is easily decomposed by concentrated hydrochloric acid, the others only with continued boiling, more readily by sulphuric acid. Ripidolite fuses with difficulty (5.5) to a grayish-yellow enamel; prochlorite becomes black and slightly magnetic. Ripidolite gives, with borax, a clear chromium-green glass, and prochlorite a glass colored by iron, which loses color on cooling. Ripidolite is monoclinic; prochlorite, hexagonal.

Leuchtenbergite, $H. = 2.5$. Colorless, white, yellowish-white. Before the blowpipe exfoliates and fuses with difficulty on the thin edges, becoming white and opaque. **Penninite**, $Mg_5AlSi_3O_{14} + 4aq$. $H. = 2.5–3$ often on edges. Color green, red, and white. With fluxes, all the vari-

eties give the reaction for iron, a
mium. **Completely decomposed**
Chloritoid, $(Fe,Mg)(AlFe)SiO_6$ +
ceptibly acted upon by hydrochlo
pletely decomposed by sulphuric aci
by ignition, $7\frac{1}{2}$ per cent.

Cerolite. (Compare Div. 5, p.
Yellowish-white. Greasy feel. H.
nition, 30 per cent. Mostly dec
chloric acid. Does not adhere to

Beauxite, $(AlFe)O$, + 2 aq. Amo
tic, grayish, reddish-brown, and red.
Loss by ignition, 20 per cent. Only
hydrochloric acid, but completely
trated phosphoric acid.

Compare *Argillite*.

Wolchonskoite. (See Div. 5, p.
Color dark-green. Boiled with pho
an emerald-green solution, which, i
retains its color, but gelatinous s
Chromite also gives the chromium r
is black, and streak yellowish-brow

Warwickite, $2 MgTiO$, + Mg_4B_6C
decomposed by sulphuric acid; ev
and moistened with alcohol, it giv
If this mass is boiled with hydroc
foil, and concentrated, the solution
with water, rose-red.

Enstatite, $MgSiO_3$; **Anthophy**
Cleavage of enstatite very perfect i
thophyllite cleaves in two directio
The former is of clove-brown o
color, with a pearly-metallic lustre;

phyllite is much less perfect. Hardness 5–5.5. **Hypersthene**, $(Mg,Fe)SiO_2$, is closely related, and cleaves at $86\frac{1}{2}°$. H.=5.5. On charcoal, before the blowpipe, gives a magnetic mass.

Tungstite, WO_3. Boiled with phosphoric acid, it gives a bluish solution, which, shaken while warm with iron filings and a little water, becomes dark-blue. Occurs in soft, earthy, yellow masses.

Scheelite, $CaWO_4$. Fusibility 5; hardness 4.5–5. The pulverized mineral, on being boiled with nitric acid, leaves a lemon-yellow residue of tungstic acid. Gives the reactions of tungstic acid (Table II.).

Cassiterite, see par. 299.

Octahedrite, **Rutile** (both dimetric, with adamantine lustre), and **Brookite**, (trimetric), TiO_2. Give the reactions of titanic acid (Table II.). On fusing the pulverized minerals with caustic potash, dissolving the fused mass in hydrochloric acid, and boiling the solution with metallic tin, it assumes a violet color, which turns to red on addition of water. Color of octahedrite, various shades of brown, passing into indigo-blue; of rutile, mostly brownish-red or red, sometimes yellowish or black; of brookite, hair-brown, yellowish, or reddish (variety *Arkansite* is iron-black). Hardness of octahedrite, 5.5–6; of rutile, 6–6.5; of brookite, 5.5–6.

Euxenite, $(Y,Fe,U)_3Ti_2Cb_2O_{12}H_2O$, and *Æschynite*, $(Ce, La, Di, Fe, Y)_3Cb_2(Ti,Th)_3O_{14}$; *Pyrochlore*, $Cb_2O_5TiO_2$, ThO_2CeO,CaO,FeO,Na_2O,F. Treated like the preceding with potash, etc., the solution, on reaching a certain degree of concentration, assumes a fine blue color, on addition of water, which changes in the air to olive-green, and gradually disappears. Æschynite swells before the blowpipe, and turns yellow or brownish. The color is

black; the powder light-brown. E
fore the blowpipe. Color brownish-
dish-brown. They have a metallic
rochlore is distinguished by its octal
brownish-red, and powder light-yel

Opal, SiO, + aq. Amorphous.
pipe yields water and becomes opa
to a clear bead, with effervescence.
with potassium hydrate, it dissolves
great extent; the solution gives a ę
with ammonium chloride. Hardness

Xenotime, $(Y,Ce)_3P_2O_8$. Color var
or flesh-red. Hardness 4–5. G.=4
phoric acid reaction, par. 94. Infu
phosphorus dissolves with great dif
glass.

(See also *Childrenite* and *Orthocla.*

Section 2. Hardness 7.

(See *Cassiterite,* *Rutile,* and *Opa*
section, whose hardness sometimes

Quartz, SiO,. The several varieti
crystal, Amethyst, Hornstone, F
etc., are infusible and unalterable b
and fuse with soda to a transparent
cence. Alone in the strongest hea
infusible. In a fine powder, fused w
is more or less soluble in water. I
monium chloride in excess gives a l
tate (*hydrated silica*). H.=7, anę
steel; G.=2.6. *Tridymite,* SiO,, cę
scopic hexagonal plates. G.=2.2–ę

Iolite, $(Mg,Fe)_2Al_5Si_5O_{18}$; **Stauro**

Si_6O_{30}, do not fuse to a transparent glass with soda. Fusibility of iolite, 5–5.5; color blue, grayish; G.=2.6. Staurolite is infusible; color brownish-red, brown; crystals often cruciform; G.=3.6.

Beryl, $Be_3AlSi_6O_{18}$; *Euclase,* $H_2Be_4AlSi_2O_{10}$; *Phenacite,* Be_2SiO_4; Zircon, $ZrSiO_4$. Hardness 7.5. Beryl and euclase turn milk-white with strong heat, and become rounded on the edges; beryl crystallizes in hexagonal prisms and possesses quite distinct basal cleavage; color usually pale-green or emerald-green. Euclase crystallizes in clinorhombic prisms, and possesses distinct cleavage in two directions at right angles to each other; color pale mountain-green, passing into blue and white. Phenacite and zircon do not change before the blowpipe, excepting that zircon becomes colorless; color red, yellow, or colorless; zircon sometimes brown or gray. Zircon, powdered, fused with potash, and boiled with hydrochloric acid, the diluted solution colors turmeric paper orange-red. If the acid solution is concentrated to crystallization, boiled with a saturated solution of potassium sulphate, a white precipitate is formed (*zirconia*). Phenacite is a little harder (8) than zircon. G. of zircon, 4.4; of the others, 2.7–3.

Topaz, $AlSi(O,F_2)_5$. H.=8; G.=3.5. Orthorhombic. Before the blowpipe infusible. Some varieties take a wine-yellow or pink tinge when heated. Fused in an open tube with salt of phosphorus, gives the fluorine reaction. With cobalt solution the pulverized mineral gives a fine blue on heating. Very slightly attacked by sulphuric acid.

Ouvarovite [Lime-chrome Garnet], $Ca_3CrSi_3O_{12}$. Emerald-green; infusible, but by ignition becomes blackish-green, and on cooling again, emerald-green. H.=7.5–8; G.=3.5. Gives with fluxes the chromium reactions (Table II.).

Spinel, (Mg,Fe)(AlFe)O, (*Pleonas*
Mg)(AlFe)O₄. Hardness 7.5–8; c
sively in octahedral crystals. Spinel
pulverized, are soluble in salt of ph
spinel, red, blue, brownish ; of pleo
ite is almost insoluble in salt of pho
color dark-green or black. *Kreith*
containing zinc and iron, slightly m
tion ; so also *Dysluite*. The specifi
three = 4.3–4.6; and of the other sp

Diamond, C. Characterized by its
passes that of corundum. H. = 10 ;

Mineral Coal, see pars. 306–308.
black ; slightly soluble in camphene.
pitchy-black ; mostly soluble in cam
see par. 309.

Simple Hydrocarbons. *Mars*
leum, mineral oils of density from
ite, Osocerite, Urpethite, Zietrisikite,
long here. *Elaterite*, see par. 311. *I*
are of the camphene series ; also
Naphthalin and *Idrialite* belong to
and probably *Aragotite.*

Oxygenated Hydrocarbons. *A*
Copalite, Ambrite, Scleretinite, Euosn
Geomyricite, are wax-like. Of those
Guyaquillite, Torbanite, Ionite, Woll
manite and *Dysodile* part of the ox;
sulphur.

Acid Oxygenated Hydrocarb
rellite (*Bog-butter*), *Succinellite, Dop;*

The oxidized minerals, arranged according to their fusibility and behavior before the blow-pipe on charcoal with sodium carbonate.

(From *Plattner's Blow-pipe Analysis.*)

a. Minerals fusible to a bead.

a. With soda yield a fluid bead :

Acmite,	Elaeolite,*	Oligoclase,
Allanite,*	Eudialyte,	Pyrasmolite,
Axinite,*	Garnet,	Sassolite,*
Boracite,*	Helvite,	Scapolite,*
Borax,*	Hydroboracite,*	Sodalite (Greenland),
Botryolite,*	Ilvaite,	Spodumene,
Crocidolite,	Labradorite,	Talc, black,
Cronstedtite,	Lapis-Lazuli,	The Zeolites.*
Datolite,*	Mica, from primitive limestone,	

b. With a little soda a bead, with more a slaggy mass :

Amblygonite,	Okenite,	Rhodonite,*
Fluorite,	Orthite,*	Sodalite,
Garnet, manganiferous,	Pectolite,	Sordawalite,
Manganese, black sili-cate (hydrous tephroite ?),	Pyrorthite,	Vesuvianite.*

c. With soda only a slag :

Amphodelite,	Iron, phosphates of sesquioxide,	Saponite,
Autunite,		Scorodite,
Brevicite,	Pharmacolite,	Tourmaline, potash,
Fahlunite,	Pharmacosiderite,	(black),
Haüynite,	Polyhalite,	Triphylite,
Heterosite,	Pyrargillite,	Wolframite.
	Pyrope,	

d. Sink with soda into the charcoal :

Celestite, Witherite.

* Denotes that the mineral fuses with intumescence, effervescence, etc.

e. Fuse with soda at first m
mass, but are decomposed by
and leave behind an infusible
into the coal:

Anhydrite,   Gay-Lussite,
Cryolite,   Glauberite,

f. Yield with soda regulin
of reducible metallic oxides a

b. Minerals which fu

a. With soda yield a fluid

Albite,   Nephelite,
Anorthite,   Orthoclase,
Emerald (beryl),   Petalite,
Euclase,*   Sodalite (1

b. With little soda a fluid

Enstatite,   Epidote,*   Hyperst

c. Yield with soda only a

Carpholite,   Pimelite,
Dichroite (iolite) blue,   Pinite,
Lazulite,*   Plumbogum
Mica,* from granite,   Pyrochlore,

d. With soda sinks into th
Barite.

e. Fuse or only swell up w
by a sufficient amount, leavin
soda sinks into the coal:

Apatite (swells up),

c. Infusible Minerals.

a. Give with soda a fluid bead :

Agalmatolite,	Hisingerite,	Quartz,
Dioptase,	Leucite,	Rutile,
Fire-clay,	Pyrophyllite,*	Sideroschisolite,
		Wolchonskite.

b. With little soda, a bead ; with more, a slaggy mass.

Cerite,	Gadolinite,*	Talc,
Chrysolite,	Phenacite,	Tourmaline* (lithia).
	Picrosomine,	

c. Yield, with soda, only a slag :

Aeschynite,*	Chrome Ochre,	Ouvarovite,
Allophane,	Chrysoberyl,	Polymignite,
Aluminite,	Cyanite,	Spinel,
Alunite,	Diaspore,	Staurolite,
Alunogen,*	Fluocerite,	Tantalite,
Andalusite,	Gahnite (a zinc coat),	Thorite,
Brucite,	Gehlenite,	Titanic Iron,
Calamine (a zinc coat),	Gibbsite,	Topaz,
Cassiterite, with much	Iron, sesquioxide and	Wörthite,
soda metallic tin,	its sulphates,	Xenotime,
Chloritoid,	Manganese, oxides,	Yttrocerite,
Chromite,	Oerstedite,	Yttrotantalite,
		Zircon.

d. Fuse or only swell up with soda, but are decomposed by a sufficient amount, and the soda sinks into the coal, leaving an infusible crust :

Alum (kalinite),	Calcite,	Magnesite,
Aragonite,	Dolomite,	Wavellite.*
Barytocalcite,	Epsomite,	

e. Sinks with the soda into the coal :
Strontianite.*

* Denotes that the mineral fuses with intumescence, effervescence, etc.

25 *

USE OF CITRIC ACID IN, ▩▩

By Prof. H. CARRINGTON

Annals of the New York Academy o

The organic acids in common
tartaric, and citric acids, possess a
minerals little short of that of hy(
some cases may advantageously rep
Citric acid is the strongest in its
trated solutions, with the addition (
or of potassium iodide, decompos
characteristic reactions.

The solution of citric acid em|
the cold, and the minerals are ver
Only a limited number of reactio

This method of examining min
bility is particularly useful in *fiel*
easy portability of the solid orga
acid may be carried in a pasteboa
in water obtained in the field.
quired, simple tests may be made
dispensable to the travelling min(

Carbonates

All carbonates (in fine powder)
cence in a strong solution of citric
siderite require to be heated.

Special reactions. Calcium carbon
solution of oxalic acid, yielding a fi
Siderite dissolves in the same (
light-yellow granular precipitate o
Witherite yields handsome feathe
Cerussite and oxalic acid yields a h

Sulphides.

Stibnite, galenite, sphalerite, pyrrhotite, and alabandite (in fine powder) are decomposed by a strong solution of citric acid in the cold, yielding H_2S more or less freely.

Bornite, bournonite, kermesite, and jamesonite act similarly on heating.

Argentite, pyrite, chalcocite, marcasite, niccolite, smaltite, chalcopyrite, ullmannite, arsenopyrite, tetrahedrite, (uraninite), millerite, linnæite, cobaltite, pyrargyrite, berthierite, tennantite, bismuthinite, stephanite, hessite, and hübnerite, are decomposed by heating with a solution of citric acid, to which solid KNO_3 (or $NaNO_3$) has been added.

Cinnabar, realgar, and orpiment resist this mixture, but are decomposed by heating with a solution of citric acid and KI (added in solid form).

Molybdenite and proustite resist both these mixtures.

Special reactions. Cinnabar dissolves quickly in a cold solution of oxalic acid, to which solid KI is added ; if the KI is not in excess, scarlet crystals of HgI_2 form abundantly.

Galenite dissolves in a mixture of $\bar{C}$ and KI, yielding lustrous yellow flaky crystals of PbI_2.

Distinguishing characteristics. Pyrite, chalcopyrite, and chalcocite are *not* decomposed by citric acid alone, while bornite and pyrrhotite are attacked.

Galenite may be detected in argentite, since the former is attacked by cold $\bar{C}$, yielding H_2S, and the latter is not attacked on boiling.

Pyrargyrite is decomposed by heating with $\bar{C} + KNO_3$, or with $\bar{C} + KI$, while proustite is not attacked.

Sundry minerals.

The following minerals dissolve in a strong cold solution of citric acid without evolution of gases : Brucite,

TABLES SHOWING THE BEHAVIOR OF CERTAIN MINERALS WITH CITRIC ACID ALONE AND WITH REAGENTS.

By Prof. H. Carrington Bolton, Ph. D.

I.

Decomposed (in fine powder) by a saturated solution of Citric Acid.

1. IN THE COLD.

A. Without evolution of Gas.	B. With liberation of CO_2.	C. With liberation of H_2S.	D. With separation of SiO_2.
Clausthalite,	Calcite, !	Stibnite,	Wollastonite,
Leucopyrite,	Dolomite,*	Galenite,	Rhodonite, !
Atacamite,	Gurhofite, !	Alabandite, !	Chrysolite,
Brucite,	Ankerite,*	Sphalerite,	Willemite, ! ‡
Gummite,	Rhodochro-	Pyrrhotite.	Nephelite,
Pyromorphite,*	site,*		Lapis-lazuli,
Mimetite,	Smithsonite,*		Chondrodite,
Triphylite,	Aragonite, !		Pectolite, ! ‡
Triplite,	Witherite, !		Laumontite, ! ‡
Vivianite, !	Strontianite, !		Chrysocolla, !
Libethenite, !	Barytocalcite, !		Calamine, ! ‡
Olivenite, !	Cerussite, !		Apophylite,
Pseudomala-	Malachite, !		Thomsonite, !
chite,	Azurite.*		Natrolite, ! ‡
Wavellite,			Mesolite, !
Pharmacosider-			Analcite,
ite, !			Chabazite,
Torbernite,			Herschelite, ‡
Autunite,			Stilbite,
Ulexite, !			Deweylite,
Cryptomor-			Prochlorite.
phite, !			
Anglesite,			
Brochantite. !			

! Completely decomposed or dissolved.   * Feebly attacked.
‡ Gelatinizes.

2. ON BOILING.

E. *Without evolution of Gas.*	F. *With liberation of CO_2.*	G. *With liberation of H_2S.*	H. *With separation of SiO_2.*
Cuprite,!	Hausmannite, †	Bornite,	Tephroite,‡
Zincite,	Pyrolusite,! †	Jamesonite,*	Ilvaite,
Melaconite,	Manganite, †	Bournonite,*	Phlogopite,*
Goethite,*	Psilomelane, †	Boulangerite,	Datolite,!‡
Limonite,*	Wad,! †	Kermesite,	Prehnite,*
Allanite (?),	Magnesite,!		Heulandite,
Apatite.*	Siderite,!		Serpentine,
Wolframite,*			Chrysotile,
Wulfenite,			Retinalite,
Crocoite,			Bastite,
Gypsum,*			Genthite,
			Gieseckite,*
			Jefferisite,
			Masonite,*
and those in A.	*and those in B.*	*and those in C.*	*and those in D.*

II.
DECOMPOSED BY BOILING WITH A SOLUTION OF CITRIC ACID ON THE ADDITION OF—

I. *Sodium Nitrate.*		K. *Potassium Iodide.*	
Silver,	Smaltite,!	Realgar,*	Pargasite,*
Mercury,	Cobaltite,!	Orpiment,*	Olivine,
Copper,	Ullmannite,!	Cinnabar,!	Almandite,
Arsenic,	Marcasite,!	Hematite,*	Pyrope,
Antimony,	Arsenopyrite,!	Menaccanite,*	Colophonite,
Bismuth,	Nagyagite,	Washingtonite,*	Epidote,
Sulphur,*	Covellite,!	Magnetite,*	
Bismuthinite,	Berthierite,!	Franklinite,	*and most of*
Domeykite,!	Pyrargyrite,	Braunite,	*those in*
Argentite,	Tetrahedrite,!	Enstatite,	*A, B, C, D,*
Hessite,	Tennantite,!	Hypersthene,	*E, F, G, H,*
Chalcocite,!	Stephanite,	Augite,	*and I.*
Tiemannite,!	Polybasite,!	Spodumene,*	
Millerite,!	Enargite,!	Hornblende,*	
Niccolite,!	Uraninite,!	Actinolite,*	
Pyrite,!	Hübnerite,		
Chalcopyrite,!	*and those in C*		
Linnæite,	*and G.*		

! Completely decomposed or dissolved
† The CO_2 evolved is derived from the citric acid.
* Feebly attacked.
‡ Gelatinizes.

III.			
L. MINERALS NOT DECOMPOSED BY THE ABOVE REAGENTS.			
Graphite,	Fibrolite,	Chromite,	Oligoclase.
Molybdenite,	Kyanite,	Chrysoberyl,	Albite.
Diopside,	Topaz,	Cassiterite,	Orthoclase.
Petalite,	Titanite,	Rutile,	Tourmaline.
Asbestus,	Staurolite,	Quartz,	Kaolin.
Beryl,	Bowenite,	Hyalite,	Ripidolite.
Zircon,	Talc,	Muscovite,	Columbite.
Vesuvianite,	Proustite,	Lepidolite,	Samarskite.
Zoisite,	Fluorite,	Wernerite,	Scheelite.
Iolite,	Cryolite,	Leucite,	Barite.
Biotite,	Corundum,	Anorthite,	Celestite.
Andalusite,	Spinel,	Labradorite,	Anhydrite.
(Two hundred species.)			

N. B.—The gases evolved are examined with lead acetate paper; the solutions with appropriate reagents.

CHAPTER

CHARACTERISTICS OF THI
ORES: THEIR BEHAVIOR
PIPE AND WITH SOLVEN

OF the physical properties of
described in this chapter, only
which serve best to discrimina
ores from each other. For a m
the student must refer to Dana
mineralogy. Among the distin
of minerals, crystalline form, has
ity stand foremost. The latter
without a balance, and will, for
less use to the practical man th
hardness.

For scales of hardness, fusibili
tallization, see pp. 217, 221.

ORES OF ANTI

200. **Stibnite.**—Gray Antimo
=4.5. 71.8 Sb. Trimetric. I
streak, and metallic lustre. Un
ture, consisting of a vast numbe
tals, sometimes side by side, som
brittle.

It fuses readily in the flame of
sometimes yields a slight sublima

ing the heat by application of the blowpipe flame, a coating is produced which after cooling is brownish-red, and which consists of a mixture of antimony trisulphide with antimony trioxide. In an open glass tube, emits sulphur dioxide and antimony fumes. On charcoal it is volatilized, covering the charcoal with antimony oxide, which, when touched with the reducing flame, disappears with a pale greenish-blue tinge.

When pure, wholly soluble in hot hydrochloric acid, with evolution of sulphuretted hydrogen; usually a residue of lead chloride is left. Partly decomposed by caustic potash; the solution, when mixed with an acid, affords a yellowish-red precipitate.

201. Berthierite. $FeSb_2S_4$. 57 Sb, 13 Fe. H.=2–3; G.=4–4.3. Trimetric. Metallic lustre less splendent than stibnite; color dark steel-gray; surface often covered with iridescent spots.

Heated in a matrass, fuses and yields a slight sublimate of sulphur; on application of a strong heat, a black sublimate of antimony sulphide is formed, which, on cooling, becomes brownish-red. In an open glass tube it behaves like the preceding ore. On charcoal, fuses easily and coats the charcoal with antimony-oxide; there remains, finally, a black slag, which is attracted by the magnet and gives with fluxes the iron reaction.

Soluble in hydrochloric acid.

202. Kermesite.—Red Antimony. $2Sb_2S_3 + Sb_2O_3$. 75.3 Sb. H.=1–1.5; G.=4.5–4.6. Monoclinic. Usually in tufts of capillary crystals of cherry-red color, with adamantine lustre. Sometimes in thin leaves, flexible and sectile. Streak brownish-red.

In a matrass, fuses readily and yields a slight yellowish-red sublimate; with strong heat, boils and gives a

26

black sublimate, which when ■■
an open tube and on charcoal, ■

It dissolves in hydrochloric aci∙
phuretted hydrogen. The powder
with caustic potash, assumes an ■
solves completely.

Minerals containing antimony: Al■
montite, Cervantite, Livingstonite, etc.

ORES OF AR■

203. Native Arsenic. As, witl
Fe, and Bi. H.$=$3.5; G.$=$5.9∙
metallic lustre; color and streak
exposure to air to dark-gray.

Heated in a matrass, sublim■
like pure arsenic. In both cases
left, which, when treated with fl∙
tions of iron, cobalt, and nickel

204. Orpiment. As_2S_3. As 61.
Trimetric. A foliaceous mineral
and streak, and resinous or pear∙

Before the blowpipe behaves l
this difference, that the sublimate
yellow and transparent.

Soluble in aqua regia and caust

205. Realgar. AsS. 70.1 As.
3.6. Monoclinic. Usually of b∙
orange-yellow color, and resinous
translucent; sectile; streak orang∙

In a matrass, fuses, boils, an∙
sublimate, after cooling, is red a
open glass tube, when carefully
mate of arsenic trioxide, sulphur ∙

charcoal, fuses readily and burns with a yellowish-white flame, emitting grayish-white fumes which possess the peculiar alliaceous odor. Subjected to the treatment described (see par. 70), a sublimate of metallic arsenic is obtained.

Not easily affected by acids, but aqua regia dissolves it with continued digestion, part of the sulphur being precipitated. A hot solution of caustic potash decomposes it, leaving a brownish-black, insoluble powder.

206. Arsenolite.—White Arsenic. As_2O_3.. $H. = 1.5$; $G. = 3.1$. Isometric. Occurs usually in minute capillary crystals of a white color and streak, and vitreous or silky lustre.

Before the blowpipe it behaves like pure arsenic trioxide (see pars. 11 and 15, and Table II., 2).

Slightly soluble in hot water; more so in water acidulated with hydrochloric acid.

Minerals containing arsenic: Arsenopyrite, Scorodite, Polybasite, Enargite, Domeykite, Whitneyite, Algodonite, Smaltite, Cobaltite, Niccolite, Pharmacosiderite, Arseniosiderite, etc.

ORES OF BISMUTH.

207. Native bismuth. Bi, with occasional traces of S, As, and Fe. $H. = 2-2.5$; $G. = 9.7$. Hexagonal. Color and streak silver-white tinged with red; lustre metallic; brittle when cold, but when hot may be laminated. Occurs foliated, granular, and arborescent; occasionally crystallized.

Before the blowpipe it behaves like pure bismuth (see pars. 17, 22).

Readily dissolved by nitric acid; the bismuth is precipitated by water.

208. Bismuthinite. Bi_2S_3. 81.25 Bi. $H. = 2$; $G. =$

6.4–6.55. Trimetric. In acicular c
of metallic lustre and lead-gray colo
or iridescent tarnish. Streak lead-gi

In a matrass, fuses and yields a slig
phur. Carefully heated in an open tub
sulphur dioxide and a coating of bis
latter may be fused, by application of
to brown drops, which, when cold,
opaque. On charcoal, fuses and b
small drops in a state of incandescer
coating of bismuth trioxide.

Soluble in nitric acid with depositio
solution gives a white precipitate with

209. Tetradymite.—Telluric Bismu
variable proportions, with sometime
61 Bi, 34.48 Te. H. = 1.–25; G.=
onal. Of pale steel-gray color, and
tre. Occurs usually in tabular crystals
the laminæ are elastic. It soils paper.

In an open glass tube it fuses readil
smoke which partly condenses, coatin
assay-piece with a white powder, in
spots; on directing the flame on this
colorless drops (tellurium dioxide), w
mate (selenium) disappears. On char
to a metallic globule, which, when tou
flame, imparts a bluish-green color to t
times gives out selenium vapors, and d
assay-piece a dark orange coating surr
distance by a white coating.

Soluble in nitric acid.

210. Bismite.—Bismuth Ochre. B.
nute quantities of Fe_2O_3, CuO, and

G.$=$4.36. Occurs usually pulverulent or earthy. Streak, straw-colored. Color greenish-yellow to grayish-white.

Before the blowpipe it behaves like pure bismuth trioxide. Soluble in nitric acid.

211. Bismutite. $2Bi_3C_3O_{10} + 9H_2O$. $89.75 Bi_2O_3$. H.$=$4-4.5; G.$=$6.9. Streak usually of a white or light-greenish color; lustre vitreous; in acicular crystallizations.

In a matrass decrepitates, yields a little water, and turns gray. On charcoal, fuses very readily and is reduced, with effervescence, to a metallic globule, covering the charcoal with a coating of bismuth trioxide. If the blast is kept up for some time, the whole of the bismuth is volatilized, and there remains a scoriaceous mass which in the reducing flame may be fused to a globule, and which with fluxes gives the indications of copper and iron. With soda it usually gives the sulphur reaction (see par. 121).

Dissolves in nitric acid with effervescence; also in hydrochloric, giving a solution of yellow color.

Minerals containing bismuth: Maldonite, Joseite, Aikinite, Chiviatite, Emplectite, Wittichenite, Eulytite, Bismutoferrite, etc.

ORES OF CHROMIUM.

212. Chromite.—Chromic Iron. $FeCrO_4$, or $(Fe,Mg,Cr)(Al,Fe,Cr)O_4$. $68 Cr_2O_3$. H.$=$5.5; G.$=$4.3-4.6. Isometric. Occurs usually massive; of iron-black or brownish-black color, with a shining and somewhat metallic lustre. Some varieties are magnetic. Streak brown.

Heated in a matrass, remains unchanged. Infusible in the forceps. After having been exposed to the reducing flame is attracted by the magnet. In borax and salt of phosphorus, slowly but completely soluble to a transparent glass, which is emerald-green after cooling. Mixed

with soda and nitre and heated ~~on pla~~
fuses and becomes yellow. With soal
reducing flame, it yields metallic iron

Concentrated acids affect it but litt
pulverized; they dissolve only a little
acid potassium sulphate, potassium ch

Minerals containing chromium: Crocoite, :
linite, Wolchonskoite, etc.

ORES OF COBAL

213. Linnæite.—Cobalt Pyrites.
H. = 5.5; G. = 4.8–5. Isometric.
bright steel-gray color, and metallic l
gray.   On charcoal, yields sulphur (
netic globule; often also arsenic fume
acid, giving a rose-red solution.

214. Smaltite.—Smaltine. (Co,Fe,
6; G. = 7.4–7.2. Isometric.   Of ti
color and metallic lustre.   Streak ■

In a matrass, usually yields, when
sublimate of metallic arsenic.   In an
fords a copious sublimate of crystalli
and sometimes emits sulphur dio■
fuses readily, with emission of copi
to a grayish-black magnetic globul
fluxes, gives the indications of iron,

With nitric acid it gives a pink sol
ide being deposited.

215. Cobaltite.—Cobaltine.   CoAs
H. = 5.5; G. = 6–6.3. Isometric.
sometimes reddish color, and meta
gray.

Unchanged in the matrass.   In ■

yields a sublimate of arsenic trioxide and vapors of sulphur dioxide. On charcoal, emits copious arsenic and sulphur fumes and fuses to a dull-black metallic globule, which is attracted by the magnet, and which, when treated with fluxes, gives the indications of cobalt and iron, and sometimes also of nickel.

Dissolves in hot nitric acid, arsenic trioxide and sulphur being deposited.

In an open glass tube sulphur dioxide is abundantly evolved, and sometimes a light sublimate of arsenic trioxide formed. On charcoal, small pieces of the mineral readily fuse to a globule, which, when cold, is covered with a black, rough crust, and which is attracted by the magnet. The pulverized mineral, after having been well calcined, dissolves in borax in the oxidizing flame to a blue, transparent bead. In a highly saturated bead of this kind, when treated on charcoal with the reducing flame, particles of metallic nickel may be seen floating about.

216. Asbolite.—Earthy Cobalt. It is a variety of Wad (see par. 273), containing sometimes a considerable quantity of cobalt oxide (40 per cent.), in combination with silica or arsenic.

With borax in the oxidizing flame, gives a dark-violet glass, which in the reducing flame becomes blue. The salt of phosphorus bead, when treated on charcoal with metallic tin, frequently exhibits the copper reaction. With soda on platinum-foil it shows the presence of manganese.

Soluble in hydrochloric acid with evolution of chlorine; the solution is usually blue, and on addition of water becomes rose-red.

Soluble in nitric acid, forming a rose-red solution, the sulphur separating out.

217. Erythrite. — Cobalt Bloom. $Co_3As_2O_8 + 8$ aq. $37.55 Co_2O_3, 38.43 As_2O_5$. H. $= 1.5$-2.5; G. $= 2.9$. Monoclinic. Usually of crimson or peach-red color; when crystallized, of pearly lustre; frequently dull and earthy, forming incrustations. Streak, pale-red.

Heated in a matrass, loses water, and the color changes to blue or green. A small crystal exposed to the inner flame fuses and colors the outer flame pale-blue. On charcoal in the reducing flame, emits arsenic fumes, and melts to a dark-gray globule of cobalt arsenide, which, with fluxes, gives the pure cobalt reactions.

Dissolves readily in acids; the solution is rose-colored; in concentrated hydrochloric acid, appears blue while hot. The pulverized mineral is partly decomposed by caustic potash; the powder assumes a bluish-gray color and the solution is sapphire-blue.

Minerals containing cobalt: Carrollite, Glaucodot, Chathamite, Skutterudite, Alloclasite, Bieberite, Roselite, etc.

ORES OF COPPER.

218. Native Copper. — Pure Copper. H. $= 2.5$-3; G. $= 8.8$. Isometric. Often twins. Of metallic lustre and copper-red color. Streak metallic, shining, ductile and malleable. Occurs usually massive or arborescent.

It fuses on charcoal to a globule, which, if the heat is sufficiently high, assumes a bright bluish-green surface; on cooling it becomes covered with a crust of black oxide. With the fluxes it gives the usual indications of copper.

It dissolves readily in nitric acid, giving a deep-blue color with ammonium hydrate.

219. Chalcocite. — Vitreous Copper. Cu_2S. 79.8 Cu.

H.=2.5–3; G.=5.5–5.8. Orthorhombic. Of a blackish, lead-gray color, often with a bluish or greenish tint on its surface. Occurs usually in compact masses, very often shining. Streak black; lustre shining.

Heated in a matrass, nothing volatile is given out. In an open tube sulphur dioxide is evolved. On charcoal, readily fuses to a globule, which boils and emits glowing drops, sulphur dioxide escaping abundantly; the outer flame is at the same time colored blue. With soda on charcoal it yields a globule of metallic copper.

Heated with nitric acid, it dissolves, leaving a residue of sulphur.

220. Chalcopyrite.—Copper Pyrite. Cu_2FeS_2. 34.6 Cu, 30.5 Fe. H.=3.5–4; G.=4.1–4.3. Tetragonal. Of a brass-yellow color and metallic lustre; on exposure to moist air it becomes iridescent on its surface. It occurs crystallized, but usually massive. Fracture conchoidal, uneven. It is easily scratched with a knife, giving a greenish-black powder.

Heated in a matrass, decrepitates, and yields sometimes a faint sublimate of sulphur, assuming at the same time a darker color or becoming iridescent. Heated in an open glass tube, sulphur dioxide is given out abundantly. On charcoal when heated it blackens, but becomes red on cooling; with continued heat it fuses to a black globule, which is attracted by the magnet; this globule is brittle, and reddish-gray in the fracture. The pulverized mineral, after roasting, gives with the fluxes the reactions of iron and copper. With soda on charcoal it is reduced; the metals are obtained in separate masses. Moistened with hydrochloric acid, it colors the flame green, even previous to fusion.

It dissolves in nitric acid, but more readily in aqua

regia, leaving a residue of sulphur.   Sa
ammonium hydrate as with other ores ol

221. Bornite.—Purple Copper.   Fe(
16.36 Fe.   H. $=3$;   G. $=5.5$.   Isometr
talline it usually takes the cubical form,
yellowish color ;   when massive, its coloi
reddish-brown ;   it speedily tarnishes, i
hues, mostly purple, blue, and red.
with a knife it gives a grayish powdei

Before the blowpipe it shows pretty
behavior as chalcopyrite.

Concentrated nitric acid dissolves it, l(
part of the sulphur behind.

222. Domeykite.— Arsenical Copper
Cu, 28.3 As.   H. $=3$-3.5 ;   G. $=7$-7.5.
ive, or disseminated ; lustre metallic ; c
steel-gray ;   streak blackish ;   fracture br

Heated in a matrass, yields a little v
mate of arsenic trioxide ; the assay-piece
white color.   In an open tube, affords a
mate of arsenic trioxide.   On charcoal,
emission of a strong alliaceous odor, to a
mass, which gives the copper reactions.

Readily soluble in nitric acid.
hydrochloric acid, metallic arsenic i
solved.

Algodonite.   Cu$_6$As.   83.5 Cu, 16.5 As
7.6.   Occurs massive ; lustre metallic ;
streak bronze ; fracture tough.   Same re;
kite.   Texture more granular and less ma
neyite.

223. Whitneyite.   Cu$_9$As.   88.4 Cu,
3.5 ;   G. $=8.3$.   Massive ; lustre metalli

streak bronze; fracture hackly. Same reactions as do-meykite and algodonite. Malleable.

224. Enargite. Cu_3AsS_4. 48.4 Cu, 19.1 As. Ortho-rhombic. H.$=3$; G.$=4$–4.3. Color grayish to iron-black; streak grayish-black, powder having a metallic lustre; brittle; easily cleavable; fracture uneven. Fuses on charcoal and gives a faint coating of arsenic, antimony, and generally zinc oxides.

Soluble in aqua regia.

225. Tetrahedrite.—Gray Copper. $4 Cu_2S + Sb_2S_3$, or with a portion of the copper replaced by Fe, Zn, Ag, Hg, and the antimony by As or Bi. H.$=3$–4.5; G.$=4.5$–5. Isometric; tetrahedral; color and streak between steel-gray and iron-black; lustre metallic.

Heated in a matrass, fuses, and finally yields a dark-red sublimate of antimony trisulphide with antimony trioxide. In an open glass tube, fuses and gives thick fumes of anti-mony (and arsenic trioxide) and sulphur dioxide. Mer-cury, when present, condenses in the upper part of the tube, forming a metallic mirror. On charcoal it fuses readily to a globule, emitting thick white fumes and sul-phur dioxide; coatings of antimony trioxide and of zinc oxide are deposited; the latter is nearer to the assay-piece, and may be tested with cobalt solution (see par. 54). To detect arsenic, see par. 71; to detect mercury, add to the finely-pulverized assay three times its weight of dry soda and treat the mixture as directed in par. 105. The pulverized mineral, after having been well roasted, gives with the fluxes the indications of iron and copper; with soda, affords metallic copper and a little iron. To detect silver, treat the mineral with pure lead and borax, as directed in par. 119.

When pulverized it is decomposed by nitric acid; the

solution has a brownish-green col
(and arsenic trioxide) and sulphu
Caustic potash effects a partial dec
mony trisulphide (and arsenic) ent
is, on addition of an acid, repree

226. Atacamite. $CuCl_2 + 3H_2($
$= 3-3.5$; $G. = 3.7$.   Trimetric.
massive-lamellar; color and stre;
bright-green, sometimes blackish-
green; translucent, subtranslucent;

Heated in a matrass, gives out v
sublimate, which, on cooling, becon
water shows acid reaction.   On char
ors the outer flame azure-blue, and
globule of metallic copper.   Two
on the charcoal, the one grayish
brownish, which, being played up
flame, change their place with an ;

Easily soluble in acids.

227. Cuprite.—Red Copper.   Cu
$3.5-4$; $G. = 5.8-6$.   Isometric;   c
also massive-granular, and sometim
a very intense, deep-red color; occ;
often exceedingly friable.   Streak ;

Heated in the forceps, fuses and
emerald-green; moistened with h
treated in the same manner, the co
charcoal it blackens, then fuses qui
a globule of metallic copper, which
covered with a coating of black ox

Dissolves in concentrated nitric ;

228. Melaconite.—Tenorite.   CuO
massive, pulverulent (*melaconite*);

(*tenorite*); also in shining, flexible scales. H.$=$3; G. $=$**6.25.** Lustre metallic; color iron-gray when in scales; dull and earthy, with a grayish-black color, and soils the fingers when massive or pulverulent.

Infusible; otherwise reactions like cuprite. Soluble in nitric and hydrochloric acids.

229. Chalcanthite.—Blue Vitriol. $CuSO_4 + 5$ aq. H. $=$**2.5**; G.$=$**2.21.** Triclinic; lustre vitreous; color various shades of blue; taste metallic and nauseous; subtransparent, subtranslucent.

Heated in a matrass, swells up, gives out water, and becomes white. On charcoal, colors the outer flame green, fuses, and affords a button of metallic copper, crusted with a coat of sulphide. After calcination, gives, with fluxes, the reactions of copper; sometimes also those of iron.

Soluble in water. A polished plate of iron introduced into the solution becomes coated with copper.

230. Olivenite.—$Cu_3As_2O_8 + H_2CuO_2$. 45.2 Cu. H. $=$3; G.$=$**4.1–4.4.** Trimetric. Crystallized, or in globular and reniform masses of indistinctly fibrous structure. Color and streak usually olive-green to brown; lustre vitreous, adamantine.

In a matrass, yields a little water. In the forceps, fuses to a globule and colors the outer flame bluish-green; the fused mass crystallizes on cooling. On charcoal, fuses, with deflagration and emission of arsenical vapors, to a metallic globule; the globule is white and somewhat brittle, and covered with a brown scoria. Fused with metallic lead, a globule of copper is obtained, and a mass of lead phosphate, which crystallizes on cooling.

Dissolves in nitric acid, also in ammonium hydrate.

231. Tyrolite.—$Cu_5As_2O_{10} + 9$ aq, with $CaCO_3$. 35 Cu. H.$=$**1–2**; G.$=$3. Trimetric. Usually reniform, mass-

ive; structure radiate foliaceou
Very sectile.   Lustre vitreous.

Heated in a matrass, decrepiti
and blackens.   On charcoal, fus
senic vapors to a gray scoriacec
nute globules of metallic coppe
When the mineral is fused on ch
soda and borax, until the coppei
duced and the slag dissolved i
solution is obtained in which the
be shown by the proper reagent

Dissolves in nitric acid with ef
monium hydrate, with a residue

232. Malachite. $CuCO_3 + H,$
3.5-4; G.= 3.7-4.   Monoclinic
shape of mammillated concretior
compact, and lustre shining; in
earthy, sometimes silky;   of a l
streak a somewhat paler green.

Heated in a matrass, gives out
On charcoal, fuses to a globule, a
per when the heat is sufficiently b
ceps, the outer flame is colored g
soda it behaves like copper oxide

It dissolves in acids with efferv
tinguished from other ores of gi
in ammonium hydrate.

233. Azurite.—Blue Malachite
55.2Cu.   H.= 3.5-4.2;   G.=
Occurs usually crystallized, or
columnar structure.   It is easil
fine blue color and streak; its
often nearly adamantine, lustre.

Before the blowpipe and with solvents it behaves like malachite.

234. Chrysocolla.—$CuSiO_3 + 2$ aq. $45.3 CuO$. $H.= 2-4$; $G.=2$. Occurs usually as an incrustation. It very much resembles malachite; its color is bluish-green, and it is remarkable for its great compactness; its surface is very smooth, giving it the appearance of an enamel or a well-fused slag. Lustre vitreous. Streak, when pure, white.

In a matrass, yields water and blackens, In the forceps infusible, coloring the outer flame intensely green. On charcoal in the oxidizing flame, blackens; in the reducing flame turns red. Salt of phosphorus and borax dissolve it with the usual indications of copper; the salt of phosphorus bead shows a cloud of undissolved silica. With soda on charcoal, affords globules of metallic copper.

It is decomposed by acids, silica remaining undissolved without gelatinization.

ORES OF GOLD, PLATINUM, AND IRIDIUM.

235. Native Gold.—Combination of Au and Ag in variable proportions, sometimes with traces of Fe and Cu. $H.=2.5-3$; $G.=15.6-19.5$. Isometric. The octahedron and dodecahedron the most common forms. Easily distinguished by its malleability, its cutting like lead, its high specific gravity, and its resistance to acids. Color and streak various shades of gold-yellow, and sometimes almost silver-white. It usually occurs in variously contorted and branched filaments, in scales, in plates, or in small, irregular masses.

On charcoal, fuses to a globule which, after cooling, has a bright metallic surface. With salt of phosphorus in the oxidizing flame, a bead is formed which opalizes on cooling, or becomes opaque and yellow, according to the amount of silver which it contains.

Resists the action of heated concentrated nitric acid; soluble only in aqua regia.

236. Sylvanite.—Graphic Tellurium. $AgAuTe_5$. H. **1.5-2**; G.—**8-8.3.** Monoclinic. Of metallic lustre and steel-gray color. Very sectile. Streak and color pure steel-gray to silver-white; sometimes yellow.

In an open glass tube yields a white sublimate which, when played upon with the flame, fuses to transparent drops. On charcoal, fuses to a dark-gray globule, depositing at the same time a white coating which, when touched with the reducing flame, disappears, tinging the flame bluish-green (see pars. 37, 60). It finally affords a light-yellow, malleable globule of metallic lustre.

Soluble in aqua regia, leaving a residue of silver chloride. The solution gives a white precipitate with water.

Minerals containing gold : Electrum, Maldonite, Nagyagite, Krennerite, Petzite, etc.

237. Native Platinum. Pt, usually combined with a little Fe, Ir, Os, Pd, Rh, and sometimes Cu and Pb. H. **4-4.5**; G. **16-19.** Isometric. Usually occurs in grains of silver-whitish or gray color; malleable and ductile.

Infusible before the blowpipe and not acted upon by fluxes. Soluble only in heated aqua regia. The solution gives a yellow granular precipitate with potassium chloride.

238. Iridosmine.—Osmiridium. Var. 1. Newjanskite. IrOs. G. **18.8-18.5.** Var. 2. Sisserskite. $IrOs_4$, or $IrOs_8$. G.—**20-21.2.** Hexagonal. Occurs usually in irregular flattened grains, of metallic lustre and tin-white color; but little malleable.

Newjanskite is infusible before the blowpipe; when fused with nitre in a matrass the characteristic osmium odor is produced. The fused mass is soluble in water; the solution gives, on addition of nitric acid, a green precipitate. The sisserskite, when strongly heated, gives off the osmium without the addition of nitre, but undergoes no further change.

ORES OF IRON.

239. Native Iron.—Fe. Usually massive, with octahedral cleavage. Color and streak iron-gray. Malleable and ductile. H.$=$4.5; G.$=$7.3–7.8. Attracted by the magnet. Occurs in grains disseminated through dolerite, basalt, and other igneous rocks; sometimes in masses. It is a constituent of nearly all meteorites, with variable quantities of Ni (from 1 to 20 per cent.), and traces of Co, Mg, Mn, Sn, Cu, Cr, Si, C, Cl, S, and P. H.$=$4.5; G.$=$7.3–7.8.

Infusible. On charcoal, with borax or salt of phosphorus, gives only the reactions of iron. To detect the presence of the other heavy metals, the assay-piece must be dissolved in aqua regia, the liquid mixed with ammonium hydrate in excess, filtered, and the ammoniacal filtrates precipitated with ammonium sulphide. The precipitate consists of the sulphides of nickel, cobalt, manganese, and copper, which may be collected on a filter and treated with borax on charcoal as described par. 99.

240. Pyrite.—Iron Pyrites. FeS_2. 46.7 Fe. H.$=$6–6.5; G.$=$4.8–5. Isometric. Occurs commonly in cubes. Usually of a brass-yellow color and metallic lustre; sometimes colored or brown by metamorphosis. By its supe-

rior hardness, not ~~yielding to the~~
sparks when struck with steel, it m
from copper pyrites.   Streak greenisl

Heated in a glass tube ~~closed at on~~
some sulphuretted hydrogen, and yi
~~sulphur~~; the residue is attracted by tl
on charcoal with the oxidizing flame
off with a blue flame and leaves red c
when treated with the fluxes, gives ｜
But slightly affected by hydrochlori
dissolves it, leaving a residue of su

241. Marcasite.—White Iron Pyrit
6.5 ; G. = 4.6–4.8.   Trimetric.   Cr
often twins.   Color usually light bi
times inclined to green or gray ;   c
radiated masses or crest-like aggrega
ish-gray.   Very liable to decompositi

Before the blowpipe it behaves like

242. Pyrrhotite.—Magnetic Pyrite
H. = 3.5–4.5;   G. = 4.4–4.7.   Hexa｜
resembles common iron pyrites, fron
guished by its inferior hardness, n
quickly tarnishing, and by being sl
the magnet.   Streak dark grayish-bl

Heated in a matrass, remains uncha
glass tube, emits sulphur dioxide, but
On charcoal in reducing flame, fuses
is covered with an uneven black co
the magnet, and which, on a surface
a yellowish crystalline structure and
the oxidizing flame it is converted

Soluble in hydrochloric acid, exc
with evolution of sulphuretted hydr

243. Arsenopyrite.—Mispickel. $FeS_2 + AsS_2$. 34.4 Fe. 68 As. H.$= 5.5–6$; G.$= 5–6.4$. Trimetric. Of metallic lustre and a silver-white to steel-gray color. Streak dark grayish-black. Brittle.

Heated in a matrass, yields first a red sublimate of arsenic sulphide, and afterward a black crystalline one of metallic arsenic; in an open glass tube yields arsenic trioxide and sulphur dioxide. On charcoal, emits copious arsenic fumes, and a coating of arsenic trioxide is deposited; then fuses to a globule which shows the properties of fused pyrrhotite. Frequently contains cobalt, the presence of which may be detected by the method described in par. 99. May be distinguished by its orthorhombic form from smaltite.

Soluble in nitric acid and aqua regia, leaving a residue of sulphur and arsenic trioxide; the latter dissolves with continued digestion.

244. Hematite.—Specular Iron. FeO_3. 70 Fe. H.$= 5.5–6.5$; G.$= 4.5–5.3$. Hexagonal. Of a dark steel-gray or iron-black color, and usually of metallic lustre; its powder and streak are red. Fracture subconchoidal, uneven.

Alone infusible; becomes magnetic after roasting, and gives the usual indications of iron with the fluxes; its powder dissolves readily heated with hydrochloric acid. Sometimes contains chromium and titanium, which may be detected by the processes given in pars. 85 and 125.

245. Menaccanite.—Titaniferous Iron. $(TiFe)_2O_3$ or Ti_2O_3, and Fe_2O_3 in various proportions. H.$= 5–6$; G.$= 5.5–6$. Hexagonal. Of iron-black color, usually in tabular crystals; slightly attracted by the magnet. It resembles magnetite.

Alone in the oxidizing flame, inf
flame it may be rounded at the edge:
salt of phosphorus in oxidizing flame
of pure iron oxide; but the salt of
when treated with the reducing flam
ish-red color, the intensity of which
amount of titanium oxide present; this
with tin on charcoal, turns violet (v.
To show conclusively the presence
method given in par. 125.

Dissolved by hydrochloric acid ar
separation of titanium oxide; som
with great difficulty, even when red
powder.

246. Magnetite.—Magnetic Iron Or
H.$= 5.5-6.5$; G.$=4.9-5.2$. Isome
iron-black, with a shining metallic or
its powder and streak are black; it i
by the magnet.

It fuses with difficulty, and gives th
iron with the fluxes; the pulverizec
completely in hydrochloric acid.

247. Franklinite. (Fe,Zn,Mn), Fe
a little SiO_2, Al_2O_3.   H.$= 5.5-6.5$; (
habit octahedral; occurs crystallized,
to compact; lustre metallic; color
dark reddish-brown; fracture conch
slightly on the magnet.

Infusible.   Dissolves in borax and
with manganese reaction. The borax
on charcoal in the reducing flame, bec
with soda on platinum-foil, gives m
with soda on charcoal, gives a fai

oxide, which becomes more distinct on addition of a mixture of borax and soda.

Dissolves completely in heated hydrochloric acid to a greenish-yellow liquid, a small amount of chlorine being evolved.

The zinc is easily obtained by neutralizing the acid solution of the mineral, acidifying with acetic acid, and adding sulphuretted hydrogen. (*Leeds.*)

It resembles magnetite, but gives the zinc coating on charcoal, and is not as strongly attracted by the magnet.

248. Limonite.—Brown Hematite. $H_6Fe_2O_9$. 59.9 Fe. H.=5-5.5; G.=3.6-4. Of a dull brownish-yellow color to black, earthy or semi-metallic in appearance, and often in mammillary or stalactitic forms. Streak yellowish-brown.

In a matrass yields water, and red sesquioxide remains; in platinum forceps, fusible on the edges; gives with borax and salt of phosphorus an iron reaction; the clayey varieties, treated with salt of phosphorus, give a cloud of undissolved silica; treated with soda and nitre on platinum-foil, the manganese reaction is almost always obtained.

249. Göthite. H_2FeO_4. 89.9 Fe_2O_3, 10.1 H_2O. Orthorhombic. In striated prisms, scales, and tabular; also fibrous, foliated, massive, stalactitic, and reniform. H. =5-5.5; G.=4-4.4. Lustre imperfect adamantine; color yellowish, reddish, and blackish-brown; streak brownish to ochre-yellow.

In the closed tube gives off water, and is changed into red sesquioxide. Behaves like hematite, with fluxes.

Soluble in hydrochloric acid.

250. Turgite. $H_2Fe_2O_7$. 94.7 Fe_2O_3, 5.3 H_2O. H.=5-

6; G.=3.56-4.14. Lustre sub-me
color reddish-black to dark-red ; br
botryoidal surface ; often lustrous, l

Heated in a closed tube, flies to
and yields water. Otherwise like

Its superior hardness, color of i
itation before the blowpipe disting
and limonite.

251. Melanterite. — Copperas.
FeO. H.=2 ; G.=1.8. Monocl
massive and pulverulent, of variou:
coming yellowish on exposure to ai
metallic.

In a matrass gives out sulphur dio:
shows acid reaction. Strongly heate
iron remains. Soluble in water.

252. Vivianite. — Blue Iron Ea
H.=1.5-2 ; G.=2.6. Monoclinic
or in reniform and globular masses,
sometimes as incrustation ; color wl
usually dirty blue ; vitreous lustre al
which often changes quickly to indi

In a matrass swells and gives ■
ceps, fuses to a steel-gray metallic
outer flame bluish-green. With fl■
of iron.

Easily soluble in hydrochloric l
With a solution of caustic potash,

Beraunite is of similar characte
foliated, columnar masses. H.■■
hyacinth-red to reddish-brown ; sl

253. Scorodite. FeAs$_2$O$_8$ + 4 aq
H.=3.5-4 ; G.=3.1-3.3. Trimet:

pale leek-green or liver-brown; lustre vitreous; streak white.

In a matrass yields pure water, and turns yellow. In the forceps, fuses to a gray scoriaceous slag of metallic lustre, coloring the outer flame pale-blue. On charcoal, emits arsenic vapors, and fuses to a gray magnetic slag of metallic lustre, which gives with fluxes the reactions of iron.

Not affected by nitric acid; forms a brown solution with hydrochloric acid; partially dissolved by ammonium hydrate, leaving a brown residue.

254. Siderite.—Spathic Iron. $FeCO_3$. 48.22 Fe. H. $=3.5-4.5$; G.$=3.7-3.9$. Hexagonal; color from grayish-yellow to reddish-brown; crystallizes in rhombohedrons, which are often curved, and are very distinctly cleavable; often massive; lustre vitreous; and streak light-brown.

Heated in a matrass, frequently decrepitates, carbon dioxide and carbon monoxide are given out, and a black iron oxide remains, which is attracted by the magnet. Alone infusible. With borax and salt of phosphorus it gives the pure iron reactions, and with soda sometimes those of manganese. Heated with strong acid, it dissolves readily, with brisk effervescence, but slowly in the cold.

Minerals containing iron: Chromite, Wolframite, Columbite, Tantalite, Dufrenite, Cacoxenite, Triphylite, Lölingite, Leucopyrite, etc.

ORES OF LEAD.

255. Galenite. PbS. 86.6 Pb. H.$=2.5-2.75$; G.$=7.25-7.7$. Isometric; color and streak lead-gray; of metallic lustre. Crystals usually cubical, with very perfect cubic cleavage. Octahedrons and twins not uncommon. It is generally argentiferous.

Heated in a matrass, sometimes
quently yields a slight white subli
open glass tube, emits sulphur di
being raised, gives a white sublim
Heated on charcoal, affords a glob
charcoal becoming at the same tin
sulphate and lead oxide. The glo
yields generally a little silver on cu
ence of antimony is ascertained as sh
par. 129 ; iron, par. 99, *b*.

It dissolves with some difficulty in
chloric acid, with evolution of sul
Very dilute nitric acid has no eff
stronger acid it is readily dissolve
nitrogen tetroxide. By fuming nitr
gia it is very violently acted upon,
sulphate or a mixture of the sulp
ride.

256. Bournonite. $2\,PbS + Cu_2S$.
$H. = 2.5-3$; $G. = 5.7-5.9$. Trimet
lized, and massive, granular, comp
color and streak steel-gray.

In a matrass, decrepitates, and
heat a dark-red sublimate. In an o
oxide is evolved, and abundant an
condense partly on the upper and
side of the tube ; the upper of anti
is volatile ; the lower is not volatil
mixture of antimony tetroxide, St
monate. On charcoal, fuses readil
and deposits a coating of antimony t
heat a coating of lead oxide is obta
globule, when treated with borax in

gives the reactions of copper, and the globule assumes the appearance of metallic copper.

Dissolves readily in nitric acid to a blue liquid, leaving a residue of antimony trioxide and sulphur. Aqua regia leaves a residue of sulphur, lead chloride, and lead antimonite; the solution gives a precipitate with water.

The following ores behave before the blowpipe in a very similar manner:

Geocronite. $Pb_5S + Sb_2S_8$. 15.9 Sb, 67.4 Pb. Sometimes with a little arsenic. Color lead-gray; granular. Trimetric.

Dufrenoysite. $2 PbS + 2 As_2S_3$. 57.18 Ph, 20.72 As. Color blackish, lead-gray; streak reddish-brown. Opaque; brittle. Trimetric.

Boulangerite. $3 PbS + Sb_2S_3$. 58.7 Pb, 23.1 Sb. Color bluish lead-gray, often spotted with red. Crystalline and granular. Orthorhombic.

Jamesonite. $2 PbS + Sb_2S_3$. 32.2 Sb, 43.7 Pb. Also containing Fe. Oxidized by nitric acid to a white powder, imparting no color to the solution. Trimetric.

Plagionite. $4 PbS + 3 Sb_2S_3$. Monoclinic.

Zinkenite. $PbS + Sb_2S_3$. Trimetric.

Meneghenite. $4 PbS + Sb_2S_3$. Monoclinic.

Those minerals in which a part of the Sb_2S_3 is substituted by As_2S_3 give on charcoal arsenical vapors, and in an open tube a crystalline sublimate.

257. Minium. $Pb_2O_4Pb = 90.66$. H.$= 2-3$; G.$= 4-6$. Pulverulent. Color vivid red mixed with yellow.

Before the blowpipe, behaves like lead oxide.

With hydrochloric acid, evolves chlorine and is converted into lead chloride. With nitric acid becomes brown.

258. Massicot.—Plumbic Ochre.
CO_2, CaO, Fe_2O_3, and SiO_2.   92.83 Pb
Trimetric; also isometric.   Massive.
between sulphur- and orpiment-yell
yellow.

Before the blowpipe, behaves like l

259. Anglesite.—Lead Vitriol.   Pb
$= 2.75-3$; $G. = 6.2$.   Trimetric.
small octahedral crystals with many f
quently in lamellar masses; of high
also massive and granular.   Fracture
brittle.

Heated in a matrass, decrepitates a
little water.   Treated on charcoal in
fuses to a clear bead, which, on cooling
with soda on charcoal, affords a globu
the soda is absorbed by the charcoa
placed on silver-foil, a strong sulph
the fluxes, gives the reactions of lead
iron or manganese may be detected b
shown in pars. 99, *d*, and 104.

It dissolves in acids only with great
out effervescence; by hydrochloric a
composed; the pulverized mineral is s
of caustic potash.

260. Crocoite.—Red Lead Ore.
$H. = 2.5-3$; $G. 5.9-6.1$.   Monoclini
in bright hyacinth-red crystals of
Streak orange.   Translucent.   Sectile

In a matrass, decrepitates; the crys
into minute pieces and assume a darke
coal fuses and becomes reduced with
ing of lead oxide is formed, and g

mium sesquioxide remains with the metallic globule. With soda on charcoal, affords a globule of metallic lead. With soda on platinum-foil, fuses to a dark-yellow mass, which becomes green in the reducing flame. With borax or sodium phosphate in the oxidizing flame, dissolved; the bead appears yellow while hot, but becomes green on cooling. Fused in a platinum spoon with from three to four parts of acid potassium sulphate, gives a dark-violet mass, which is greenish-white when cold.

261. Vauquelinite. $Pb_2CuCr_2O_9$. 56.4 Pb, 8.6 Cu. H. $= 2.5-3$; G. $= 5.5-5.7$. Monoclinic. Occurs usually in minute crystals, or in reniform or globular masses. Color dark-green to brown, sometimes nearly black. Lustre adamantine to resinous; streak greenish to brownish.

On charcoal, fuses with effervescence to a gray, sub-metallic globule; where the mass is in contact with the coal, small globules of lead make their appearance; in the reducing flame, a coating of lead oxide is formed. With borax or sodium phosphate in the oxidizing flame, clear green beads are obtained, which remain green on cooling, but which, on application of the reducing flame, become red and opaque; this reaction appears most distinctly on charcoal with tin. With soda on platinum wire in the oxidizing flame, dissolves to a transparent green bead, which on cooling becomes yellow and opaque; on treating the bead with a few drops of water, a yellow solution is obtained, in which the presence of chromic acid may be proved, as described in par. 68. With soda on charcoal, is completely decomposed; on treating the reduced metals with boric acid on charcoal (see par. 88), a globule of metallic copper is obtained.

Partly soluble in nitric acid to a dark-green liquid; the residue is yellow.

262. Wulfenite.—Yellow Lead Ore.
times with a little Cr.   57. Pb.   H. = 2
Dimetric.   Crystallized or granularly n
herent.   Color usually wax-color, pass
yellow.   Streak white.

In a matrass, decrepitates and becomes
On charcoal, fuses and is partly absor
while metallic lead and a coating of le
posited.   With borax or sodium phosp
wire, gives the reactions of molybdic ac
18).   With soda on charcoal, affords a
lic lead.   Fused with acid potassium s
tinum spoon, a yellowish mass is obt
comes white on cooling; treated wit
and a piece of metallic zinc placed in
liquid assumes a blue color.   Moisten
acid, heated in the platinum spoon or o
fumes escape, allowed to cool, and the
it changes to a deep-blue color.

Dissolves in concentrated hydrochlor
liquid, leaving a residue of lead chlori
ized mineral is decomposed on being di
acid; a yellowish-white residue is left
blue when exposed to air in thin laye

263. Pyromorphite. $3 Pb_3(PO_4)_2 + 1$
quently the P is replaced by As and the
Pb.   H. = 3.5-4;   G. = 6.5-7.   Hexa
often in globular masses with a column
reniform, fibrous, and granular.   Lu
Streak slightly yellow.   Color green, ye
of different shades; also white.

Heated in a matrass, sometimes decrep
with continued heat, a faint white and

of lead chloride. Heated in the platinum-pointed forceps, fuses readily and colors the outer flame bluish-green; if the amount of phosphoric acid is not too small, the edges of the flame will appear green. With salt of phosphorus and copper oxide, gives the reaction for chlorine (par. 82). On charcoal in the oxidizing flame, fuses to a globule, which, on cooling, assumes a polyhedral form and a dark color; in the reducing flame, yields a coating of lead oxide, and the globule, on cooling, assumes dodecahedral facets of pearly lustre. With magnesium wire, gives the reaction for phosphoric acid (par. 110). With soda on charcoal, affords metallic lead. When a portion of the phosphorus is replaced by arsenic, it is readily detected by the odor when treated with soda on charcoal (par. 33). Also a part of the lead is replaced by calcium, as in the brown varieties polysphærite, miesite, and nussierite, while some of the PbCl is replaced by calcium fluoride, thus diminishing the amount of lead.

Soluble in nitric acid and solution of caustic potash.

264. Plumbo-Gummite. Contains $Al_2O_3, Pb, H_2O, P_2O_5$. H.=4-5; G.=4.8-6.4. In reniform or globular masses, with a columnar structure; also compact, massive. Of resinous lustre; color white, grayish-green, reddish-yellow, but usually yellowish-brown; resembling gum-arabic in appearance. Streak colorless.

In a matrass, decrepitates and gives out water. In the forceps, intumesces and colors the outer flame azure-blue. On charcoal, intumesces, becomes white and opaque, and fuses but imperfectly, depositing a faint white coating of lead chloride. In small quantities, soluble in borax and salt of phosphorus to clear beads. With soda on charcoal, minute globules of metallic lead are obtained.

Treated with cobalt solution, assumes
Soluble in nitric acid.   The soluti
monium molybdate a yellow precipi

265. Cerussite.—White Lead Ore.
H.=3-3.5; G.=6.4.   Trimetric.
massive, in prismatic needles, or in
Rarely fibrous.   Color mostly white
Streak colorless.

When heated in a matrass decrepi
low; carbon dioxide is given out.   ]
alone, is reduced to a metallic bead.   '
dissolves with effervescence and give
pure lead oxide (see Table II., 15)
and with effervescence in dilute nitric
chloric acid, leaves a residue of lead
in a solution of caustic potash.

266. Leadhillite. $PbOSO_4 + 3 Pb$
— 2.5; G.==6.2-6.5.   Trimetric. O(
crystals of pearly or resinous lustre.
ing into yellow, green, or gray.   Stre
On charcoal, intumesces slightly, b
white again on cooling; with greater
to metallic lead.

Dissolves in nitric acid with effer
residue of lead sulphate.   **Lanarkite**

267. Phosgenite. $PbCl_2 + PbCO_3$,
2.75-3; G.=6-6.3.   Dimetric.   Fo
amantine lustre, of white, gray, or yel
white.   Transparent and translucen
tile.

In a matrass, decrepitates slightly a
darker yellow.   On charcoal, fuses r
vapors, becomes reduced to metallic

white coating of lead chloride and a yellow coating of oxide. With salt of phosphorus and copper oxide gives the chlorine reaction.

Dissolves in nitric acid with effervescence.

Minerals containing lead: Clausthalite, Mendipite, Caledonite, Mimetite, Vanadinite, Melanochroite, Stolzite, etc.

ORES OF MANGANESE.

268. Pyrolusite.—Black Oxide of Manganese. MnO_2. 63.3 Mn. H.$= 2$–2.5; G.$= 4.8$. Trimetric. Of black or steel-gray color and little lustre; powder black; sometimes of columnar structure. Streak black or bluish-black; sometimes sub-metallic. It is distinguished from psilomelane by its inferior hardness and being usually crystalline.

In a matrass, usually yields a little water; when heated to redness, oxygen is evolved. Alone infusible, but turning reddish-brown when the temperature is sufficiently high. Soluble in borax and salt of phosphorus with the usual manganese reactions; gives frequently the indications of iron.

Soluble in hydrochloric acid with disengagement of chlorine.

269. Hausmannite. Mn_3O_4. 72.1 Mn. H.$= 5$–5.5; G.$= 4.7$. Dimetric. Crystallized or granular particles, strongly coherent. Color brownish-black; streak chestnut-brown.

Before the blowpipe and with hydrochloric acid behaves like the preceding ore.

270. Braunite. $2(2MnO,MnO_2) + MnO_2,SiO_2$. H.$= 6$–$6.5$; G.$= 4.7$–$4.8$. Dimetric. Occurs crystallized or massive; color and streak brownish-black.

In a matrass *does not give any water;* behaves other-

wise like pyrolusite. Di............
disengagement of chlorine, leavi
of silica. Distinguished from the
ganese by its superior hardness.

271. Manganite. H₂MnO₄. H
$=4$; G.$=4.2$-4.4. Trimetric;
ular; often stalactitic; lustre su
steel-gray to iron-black; streak r
black; opaque; fracture uneven

In the closed tube yields wat
braunite.

272. Psilomelane. Compositio
tially MnO₂, with BaO, or K₂O ar
$=3.7$-4.3. Massive, botryoidal,
iron-black to steel-gray; streak l

In a matrass it usually *yields co*
solvents it behaves like pyrolusite

273. Wad.—Bog Manganese.
H₂O; and also often contains F
etc. H.$=0.5$-6; G.$=3$-4.2.
compact; of a dull-black color.

Varieties :

a. Bog Manganese, manganesi
b. Asbolite, cobaltiferous.
c. Lampadite, cupriferous.
Before the blowpipe—
a. Behaves like psilomelane.
b. Gives a blue bead with salt o
heated in the reducing flame on
of Sn, sometimes gives a copper

c. Gives similar reactions to
eties the manganese reaction wi
chlorine when treated with hyd

274. Rhodochrosite.—Dialogite. $MnCO_3$; the Mn often replaced in part by Ca, Mg, Fe, or Co. H.$=3.5$–4.5; G.$=3.4$–3.7. Hexagonal. Occurs crystallized or in globular masses of columnar structure; also massive; color shades of rose-red to brownish-red; streak white; lustre vitreous and inclined to pearly; translucent; sub-translucent.

In a matrass, some varieties give a little water, and decrepitate violently. Infusible. When heated in the reducing flame, does not become magnetic. Dissolves in fluxes with effervescence, and gives usually the reaction of manganese and iron.

The pulverized mineral is little affected by hydrochloric acid in the cold; on heating, dissolves with effervescence.

275. Rhodonite. $MnSiO_3$. 54.1 MnO, 45.9 SiO_2. H.$=5.5$–6.5; G.$=3.4$–3.7. Triclinic; usually massive; lustre vitreous; brownish-red, flesh-red, yellowish-red, or black on the surface from exposure; streak uncolored.

When heated becomes dark-brown, and gives to borax a deep violet while hot and reddish-brown when cold.

Resembles red feldspar, but differs in specific gravity, blackening on exposure, and coloring the borax bead.

Minerals containing manganese: Franklinite, Wolframite, Alabandite, Hauerite, Chalcophanite, Lithiopholite, Triphylite, Triplite, Dickinsonite, Reddingite, Fairfieldite, Triploidite, etc.

ORES OF MERCURY.

276. Native Mercury. Hg; sometimes containing a little Ag. G.$=13.5$. Metallic globules of a tin-white color.

Heated in a matrass, is converted into vapor, which condenses in the neck of the matrass to small metallic globules.

Dissolves readily in nitric acid.

Amalgam. AgHg,64.9 Hg, and also Ag$_2$Hg$_3$. 73.5 Hg. H.=3-3.5; G.=13.5-14. Isometric. Occurs crystallized and massive. Color and streak silver-white; opaque.

In a matrass, boils, gives a sublimate of metallic mercury, and leaves a spongy residue of silver, which on charcoal fuses readily to a globule.

Dissolves readily in nitric acid.

Arquerite. Ag$_{12}$Hg. 13.4 Hg. G.=10.8. Isometric. In regular octahedrons; also in grains, small masses, and sometimes dendritic. In color, lustre, and ductility like native silver, but softer.

277. Cinnabar. HgS. H.=2-2.5; G.=8.98. Hexagonal; color various shades of red, from cochineal-red to dark brownish-red; powder always bright red. It occurs in very small flattened crystals, or granularly massive. Streak scarlet, subtransparent to opaque.

Heated in a matrass, is volatilized, and condenses to a black sublimate, which by friction sometimes assumes a red color. Mixed with soda, yields, on heating, globules of metallic mercury. In an open glass tube is partially decomposed into metallic mercury and sulphur dioxide. On charcoal it is, when pure, wholly volatilized.

Nitric acid and hydrochloric acid have no visible effect on it. Aqua regia dissolves it, part of the sulphur being precipitated. Insoluble in caustic potash.

278. Calomel.—Horn Quicksilver. HgCl. 84.9 Hg. H.=1-2; G.=6.48. Dimetric. Occurs usually in distinct crystals, or crystalline coats, of adamantine lustre and yellowish-gray color. Translucent; streak pale yellowish-white.

In a matrass, yields a white sublimate of mercurous chloride. Mixed with soda and heated in a matrass, affords globules of metallic mercury. On charcoal, completely volatilizes, giving a white coating. Shows the chlorine reaction when treated as described in par. 82.

Treated with boiling hydrochloric acid, is partly dissolved, and becomes gray. Not affected by nitric acid; dissolved by aqua regia. With a solution of alkali, becomes black.

Minerals containing mercury: Metacinnabarite, Tiemannite, Coloradoite, Magnolite, etc.

ORES OF NICKEL.

279. Millerite. — Capillary Pyrites. NiS. 64.4 Ni. H. = 3–3.5 ; G. = 4.6–5.6. Hexagonal. Occurs usually in delicate capillary crystals of brass-yellow to bronze-yellow color, often with gray iridescent tarnish, and metallic lustre. Streak bright. Brittle.

In an open glass tube evolves sulphur dioxide. On charcoal, fuses with emission of sparks to a metallic globule which is attracted by the magnet. The calcined mineral gives with fluxes the indications of nickel oxide, and sometimes also those of cobalt oxide.

By heated concentrated nitric acid it is but little affected, but its color is changed to gray. By aqua regia it is wholly dissolved.

280. Niccolite. — Copper Nickel. NiAs, or Ni_3As_3. 43.6 Ni, 56.4 As. Sometimes part of the As is replaced by antimony. H. = 5–5.5 ; G. = 7.3–7.6. Hexagonal. Usually massive ; of copper-red color, with a gray tarnish and metallic lustre ; very brittle. Streak dark-brown.

In a matrass affords a very slight sublimate of arsenic trioxide. In an open glass tube yields a copious sublimate of arsenic trioxide, and usually a little sulphur

dioxide; the assay-piece assumes at th
lowish-green color and crumbles to p
coal, emits arsenic fumes and fuses to
tle globule, which, when treated with b
ing flame, imparts usually to the flux
and cobalt. Sometimes a faint coatin
deposited on the charcoal.

Dissolves almost completely in concer
the solution has a green color; on co
oxide separates.   Readily dissolved b

281. Gersdorffite.—Nickel Glance. N
Ni, 45.5 As. H. = 5.5; G. = 5.6–6.9.
hedral.   Of silver-white or steel-gray c
lustre.   Streak grayish-black.

In a matrass, decrepitates violently,
lowish-brown sublimate of arsenic sulp
glass tube, emits arsenic trioxide and
On charcoal, fuses with emission of
nic fumes to a globule, which, when
in the reducing flame, gives the ▰▰
cobalt.   After having removed these ▰
maining globule exhibits with the flux
pure nickel oxide.

Partly decomposed by nitric acid, ▰
tion, sulphur and arsenic trioxide bei▰

282. Ullmannite.—Nickeliferous G▰
S. + NiSh,. 27.7 Ni. Arsenic is some
= 5–5.5; G. = 6.2–6.5. Isometric. ▰
the preceding ore in its physical prope
steel-gray.

In a matrass, yields a slight white,
open glass tube, emits copious antimo
phur dioxide.   On charcoal in the re▰

o a globule and coats the charcoal with antimony tri-xide ; sometimes the odor of arsenic is observable. The melted globule, when treated with borax, frequently exhibits the reactions of iron and cobalt besides those f nickel.

It is violently acted upon by concentrated nitric acid, orming a green solution, sulphur, antimony, and arsenic rioxides being precipitated. Aqua regia dissolves it, the ulphur separating out.

283. Annabergite. $Ni_3As_2O_8 + 8$ aq. 29.2 Ni. Mono-linic. Soft, earthy. In capillary crystals, also massive and disseminated. Color fine apple-green. Streak some-what lighter.

In a matrass, yields water and darkens in color. In he forceps, fuses easily and colors the outer flame light-blue. On charcoal in the reducing flame, fuses with emission of arsenic vapor to a blackish-gray globule; when treated with borax the globule gives the reactions of nickel, sometimes also those of iron and cobalt, which it always contains.

Soluble in acids, giving a green solution.

284. Zaratite.—Emerald Nickel. $Ni_3CO_5 + 6$ aq. 59.3 NiO. H.=3-3.2; G.= 2.5-2.7. Usually forms incrusta-tions of emerald-green color and vitreous lustre. Streak pale-green.

In a matrass, loses already at 212° a considerable amount of water, and blackens. In borax and salt of phosphorus, dissolves with effervescence, exhibiting the characteristic nickel reactions.

Dissolves easily in heated dilute hydrochloric acid with effervescence.

285. Genthite. $H_4(Ni,Mg)_4Si_3O_{12}$. Amorphous. In-crusted with a delicate stalactitic surface. H.=3-4.

Some specimens very soft, and, if placed in water, crumble to pieces. G.=2.409. Lustre resinous; color yellowish or greenish. Streak greenish-white. Translucent to opaque.

In the closed tube, blackens and yields water. Infusible before the blowpipe. Gives a violet bead in the oxidizing flame, gray in the reducing flame.

Decomposed by hydrochloric acid without gelatinizing.

Minerals containing nickel: Beyrichite, Breithauptite, Morenosite, etc.

ORES OF SILVER.

286. Native Silver.—Pure silver, associated with gold, copper, and sometimes platinum, antimony, bismuth, and mercury. H.=2.5-3; G.=10-11. Isometric; twins. Color silver-white; lustre metallic; ductile and malleable. Occurs usually in twisted filaments, or arborescent; sometimes in plates or massive.

On charcoal, fuses easily to a globule, which assumes a bright surface, and shows after cooling a silver-white color. Foreign metals are detected by the methods given in pars. 117-119.

It dissolves in nitric acid, and is again deposited by a plate of copper.

287. Argentite.—Silver Glance. Ag_2S. 87.1 Ag. H. =2-2.5; G.=7. Isometric. Color blackish, lead-gray; lustre metallic. It is easily distinguished from other minerals of the same color by being cut with a knife like lead. Malleable.

On charcoal in the oxidizing flame, intumesces, gives out sulphur dioxide, and finally yields a globule of metallic silver.

Soluble in dilute nitric acid, leaving a residue of sulphur.

Jalpaite. (Ag,Cu,)S. Isometric. A cupriferous silver glance from Mexico. Color blackish lead-gray. Malleable.

Acanthite. Ag,S. Trimetric. Reactions the same as for argentite, and differs only in crystalline form.

288. Stromeyerite.—Argentiferous Sulphide of Copper. Cu,S + Ag,S. 53 Ag, 31.2 Cu. H.=2.5–3; G.=6.2–6.3. Trimetric. Occurs usually in small, compact masses. Lustre metallic; color dark steel-gray; streak gray, shining.

In a matrass, fuses easily and gives sometimes a little sulphur. In an open tube, fuses to a globule and gives off sulphur dioxide. On charcoal, fuses to a gray metallic globule, which is somewhat malleable; with fluxes the globule gives the reactions of copper, sometimes also those of iron; on a cupel with lead, affords a globule of silver.

Dissolves in nitric acid, leaving a residue of sulphur.

289. Dyscrasite.—Antimonial Silver. Ag,Sb and other proportions. 78 aq. H.=3.5–4; G.=9.4–9.8. Trimetric. Occurs crystalline or massive; granular. Lustre metallic; color and streak silver-white, also tin-white.

On charcoal, fuses readily to a gray, non-ductile globule, and coats the charcoal with antimony trioxide. With continued heat the globule assumes the appearance of pure silver and the coating becomes reddish.

290. Pyrargyrite.—Ruby Silver Ore, 3 Ag,S + Sb,S,. 59.8 Ag. H.=2–2.5; G.=5.7–5.9. Hexagonal. Color dark-red to black, giving a cochineal-red powder. Crystallizes in hexagonal prisms. Streak cochineal-red. Lustre metallic-adamantine.

In a matrass, fuses very readily, and yields with continued heat a sublimate of antimony trisulphide. In an open glass tube, gives antimony fumes and sulphur dioxide. On charcoal, fuses readily and deposits a coating of antimony trioxide, being converted into silver sulphide; if for a long time exposed to the oxidizing flame, or, when mixed with soda, in the reducing flame, affords a globule of metallic silver.

Part of the Sb_2S_3 is sometimes substituted by As_2S_3; it then gives out arsenic fumes when mixed with soda and heated in the reducing flame on charcoal.

The pulverized mineral, when heated with nitric acid, turns black, and is ultimately dissolved, leaving a residue of sulphur and antimony trioxide. Caustic potash also blackens it and affects partial solution, from which acids precipitate antimony trisulphide.

291. Proustite.—Light-red Silver Ore. $3Ag_2S + As_2S_3$. 65.5 Ag. H. = 2-2.5; G. = 5.4-5.5. Hexagonal. Very much resembles the dark-red silver ore, but is of a somewhat lighter color. Lustre adamantine.

Before the blowpipe and to solvents, behaves like the preceding, excepting it gives off arsenic fumes instead of antimony trioxide. The solution in caustic potash deposits a yellow precipitate when neutralized with acids.

292. Stephanite. — Brittle Silver Ore. $5Ag_2S,Sb_2S_3$. 68.5 Ag. H. = 2-2.5; G. = 6.2. Trimetric. Of metallic lustre and iron-black color and streak; it is very brittle and fragile.

In a matrass, decrepitates, then fuses, and ultimately yields a faint sublimate of antimony trisulphide. On charcoal, fuses very readily, and coats the charcoal with antimony trioxide. If the blast with the oxidizing flame is kept up for a sufficient time, the coating assumes a red color and a globule of metallic silver is obtained. Contains frequently copper and iron, which may be detected by the process described in par. 88. If arsenic is present, it gives in the open tube a crystalline sublimate of arsenic trioxide.

In dilute heated nitric acid it dissolves, excepting the sulphur and antimony trioxide; the solution becomes milky on addition of water. Partially dissolved by a boiling solution of caustic potash.

293. Polybasite. $9\,Ag_2S + Sb_2S_3$. 75.5 Ag. H.$=2$-3; G.$=6.2$. Trimetric. Occurs usually in short tabular prisms or massive. Lustre metallic; color and streak iron-black.

In a matrass, fuses very readily, but gives nothing volatile. In an open tube, gives sulphur dioxide and antimony fumes; the sublimate sometimes contains crystals of arsenic trioxide. On charcoal, gives a coating of antimony trioxide; with continued heat, gives a bright metallic globule, which, on cooling, becomes black on its surface; sometimes a faint coating of zinc oxide is deposited; the metallic globule affords with fluxes the reactions of silver and copper.

With acids, behaves like bournonite.

294. Cerargyrite.—Horn Silver. AgCl. 75.3 Ag. H. $=1$-1.5; G.$=5.5$. Isometric. Remarkable for its pearl-gray or greenish color, its semi-transparency, resinous lustre, and more especially for its softness, which is so great as to allow it to be marked by the nail. It turns brown on exposure to air. When rubbed with a moistened plate of zinc or iron, the latter becomes covered with a coating of silver. The streak is shining.

It fuses in a candle-flame. On charcoal, is easily reduced, especially when mixed with soda. Mixed with copper oxide and heated on charcoal in the reducing flame, copper chloride is formed, which colors the flame azure-blue (see par. 82).

29 *

Insoluble in water and nitric acid
ammonium hydrate. Partially dec
solution of caustic potash.

295. Bromyrite.—Silver Bromide
·H.= 2-3 ; G.= 5.8-6. Isometric.
small concretions. Lustre splende
green or green. Sectile.

Before the blowpipe on coal en
and yields a globule of silver. Fusec
sulphate in a matrass, gives off yel
of bromine. The globule while ho
yellow when cold. Insoluble in ni
soluble in ammonium hydrate.

296. Embolite.—Chloro-Bromide
AgCl in varying proportions. 61 t
1.5 ; G.= 5.3-5.8. Isometric. Cr
Lustre resinous; color various shad
yellow.

On charcoal, fuses readily, evolve
bromine, and affords a globule of m
soda on charcoal, reduced ; on dis
alkaline mass which has passed into
the solution to dryness, and treating
potassium sulphate as described it
vapors are given out; the bead whi
and yellow when cold. Fused wi
charcoal in the reducing flame, co
greenish, then blue (see par. 78).

297. Iodyrite. —Silver Iodide. Ag
G.= 5.7. Hexagonal. Soft. Occ
thin plates with a lamellar structure
low to yellowish-green. Lustre resi

On charcoal, fuses readily, colors t

and affords a globule of silver. In a matrass with acid potassium sulphate, gives off iodine vapors, and fuses to a very dark, almost black, globule.

Tocornalite. $AgI + HgI$. Amorphous. Color pale-yellow.

Minerals containing silver: Native Amalgam, Hessite, Petzite, Sylvanite, Miargyrite, Freieslebenite, Argentiferous Tetrahedrite, Galenite, etc.

ORES OF TIN.

298. Stannite.—Tin Pyrites. $(Cu,Fe,Zn,Sn)S$. 26 Sn. $H.=4$; $G.=4.3-4.5$. Probably dimetric and hemi-hedral. Of steel-gray or iron-black color and metallic lustre. Occurs usually massive, granular, and dissemi-nated. Streak blackish.

In an open glass tube, yields sulphur dioxide and tin oxide, which collect close to the assay-piece, and which cannot be volatilized by heat. On charcoal in reducing flame, fuses to a black scoriaceous globule; in the oxi-dizing flame, gives out sulphur dioxide and becomes cov-ered with tin oxide. When well calcined by the alternate application of the oxidizing flame and the reducing flame, gives with borax the indications of Fe and Cu. With soda and borax, yields a globule of impure cop-per.

Decomposed by nitric acid, a blue solution is obtained, and a mixture of sulphur and tin oxide remains undis-solved.

299. Cassiterite.—Tin Ore. SnO_2. 78.67 Sn. $H.=6-7$; $G.=6.3-7.1$. Dimetric. It occurs crystallized in square prisms terminated by more or less complicated pyramids; re-entrant angles are so frequent that they are to a certain extent characteristic; also massive, and in small mammillated masses of fibrous texture, hence

tected by calcining the mineral in the oxidizing flame and **treating** the residue with borax.

The pulverized mineral dissolves in nitric acid, leaving **a** residue of sulphur.

301. Zincite.—Red Zinc Ore. ZnO, containing Mn. **80.26 Zn.** H.$=4$-4.5; G.$=5.4$-5.5. Hexagonal. Of **a** deep-red color and high lustre; of distinctly foliated **structure** and orange-yellow streak.

Infusible alone. Dissolved by borax in the oxidizing **flame** with manganese reaction. With soda on charcoal **deposits** a copious coating of zinc oxide.

Soluble in nitric acid without effervescence; in hydrochloric acid with evolution of chlorine.

302. Smithsonite.—Zinc Carbonate. $ZnCO_3$. 52 Zn. H.$=5$; G.$=4$-4.5. Hexagonal. Of vitreous lustre, and white, grayish, or brownish color and streak; semi-transparent or opaque. Often stalactitic or mammillary.

Heated in a matrass, loses carbon dioxide, and, if pure, appears after cooling enamel-white. The ZnO is often to a large extent substituted by FeO, MnO, CdO, PbO, MgO, CaO; it then, after cooling, frequently assumes a dark color and gives with fluxes the indications of iron and manganese. Mixed with soda and exposed to the reducing flame, it is decomposed, and zinc oxide deposited on the charcoal, which may be tested with cobalt solution. If the temperature is raised sufficiently high, a zinc flame is sometimes observable. The coating is at first dark-yellow, or reddish when cadmium is present.

It readily dissolves in acid with effervescence; also in caustic potash.

303. Willemite.—Anhydrous Zinc Silicate. Zn_2SiO_4, and often containing a little Mn, Fe, Ca, and Mg. 72.9

Zn. H.$=$5.5; G.$=$3.89–4.27. H(
vitreo-resinous; weak. Color whitish
when purest; green to dark-brown wh(
uncolored. Transparent to opaque.

Before the blowpipe in the forceps
with difficulty to a white enamel; the \
Jersey fuse from 3.5 to 4. The pow
coal in the reducing flame, gives a coa
hot, and white on cooling, which, moi
solution and treated in the oxidizing
bright-green. With soda the coating
tained. Decomposed by hydrochloric
tion of gelatinous silica.

304. Calamine.—Hydrous Zinc Silica
67.5 ZnO, with sometimes a little lead.
—3.1–3.9. Trimetric. It closely rese
cal characters the preceding ore. It b(
heat; the smallest fragment heated ;
stances.

Infusible in the forceps. In a matras
turns milk-white. With borax dissolve:
glass, which cannot be made opaque b;
solves in salt of phosphorus to a transp
becomes opaque on cooling, and in w
saturated, clouds of silica are observabl(
soda on charcoal, swells and affords wit
ing of oxide of zinc. With cobalt s(
green color, which, when the heat is ra
fine light blue on the fused edges.

It is readily decomposed by acids, \
gelatinous silica. Dissolved by a stron
tic potash.

Minerals containing zinc: Hydrozincite, Aut

CARBONACEOUS COMPOUNDS.

305. Graphite.—Plumbago. C, with often a little iron sesquioxide *mixed* with it. Hexagonal. In flat, six-sided tables. H.= 1–2 ; G.= 2–2.2. Lustre metallic ; streak black and shining ; color iron-black to dark steel-gray ; opaque ; sectile ; marks paper ; thin ; laminæ flexible ; feel greasy.

It occurs also foliated, columnar, radiated, scaly, granular, and massive.

At a high temperature it burns without flame or smoke, leaving usually some red iron oxide. Before the blowpipe, infusible ; fused with nitre in a platinum spoon, deflagrates, converting the reagent into potassium carbonate, which effervesces with acids. Unaffected by acids.

306. Anthracite. C (from 80 to 95 per cent.), with a small percentage of SiO_2, Al_2O_3, and FeO_3. H.= 2–2.5 ; G.= 1.3–1.8. Lustre bright, often sub-metallic ; color iron-black, frequently iridescent ; fracture conchoidal.

In a matrass, gives usually a little water, but no empyreumatic oil. Heated on platinum foil in the oxidizing flame, is slowly consumed without flame, leaving a small quantity of ash, which consists of SiO_2, AlO_3, and more or less of FeO_3. Does not color a boiling solution of caustic potash.

307. Bituminous Coal. C,H,O, in variable proportions. The bituminous matter contains from 76 to 90 per cent. of carbon ; the earthy impurities consist principally of SiO_2, AlO_3, and CaO; contains frequently a small amount of N and FeS_2. Softer than anthracite. G. = 1.2–1.5. Less highly lustrous than the preceding, and of a more purely black or brownish-black color.

In a matrass, some varieties soften and cake (*caking*

coal), while others are entirely
decomposed, evolve combustible
oils, and leave a residue of me
(coke), which behaves like anth
burns with a luminous flame an
ing an earthy residue.

Boiled with a solution of cau
imparts to these solvents no col

308. Brown Coal. Composit
tuminous coal, but the organic
from 60 to 75 per cent. of carb
ties bears sometimes a close r
ing. Some varieties show disti
(*lignite*).

In a matrass, infusible, bu
evolves combustible gases, em
acid reaction, and a peculiar
a residue which consists of c
amount of ash. On platinum
flame and emission of a pecul

Boiled with a solution of c
liquid brown.

309. Asphaltum. C, H, O,
with about 75 per cent. of c
black or brownish-black color

Fuses at about 100° C., and
and emission of a thick smoke
consists essentially of SiO_2, Al
trass, gives empyreumatic oil,
combustible gases, and leaves

Treated with boiling ether,

color the liquid, or imparts at the most a pale-yellow color (distinction from brown coal).

310. Succinite.—Amber. C,H,O. H.$=$2–2.5; G.$=$ 1.065–1.081. It occurs in irregular masses, without cleavage; lustre resinous; color yellow; sometimes reddish, brownish, and whitish; often clouded; streak white; transparent; translucent; tasteless; electric on friction; fuses at 287° C., but without becoming a flowing liquid.

It consists of succinic acid, resins, an ethereal oil, and succinite proper, an insoluble substance.

Amber fuses with some difficulty in the matrass, yielding water, empyreumatic oil, gases, succinic acid, and a residue of amber resin. It burns with a yellow flame, emitting an agreeable odor, and leaving a black, shining, carbonaceous mass.

311. Elaterite.—Elastic Bitumen. C and H. G.$=$ 0.905–1.223. Soft, elastic, like India-rubber, but sometimes hard and brittle. Color dark-brown; subtranslucent. Occurs in compact, reniform, or fungoid masses. Usually with a peculiar suffocating odor.

Burns in the flame of a candle, and gives empyreumatic products when fused in a matrass.

312. Ozocerite. G.$=$0.85–0.90. In appearance and consistency similar to wax or spermaceti. Color white, yellowish, brown, and leek-green; translucent; feel greasy; wax-like odor; fusibility 56° to 63° C. It has been obtained by destructive distillation from mineral coal, peat, petroleum, etc.

HYDROCARBON COMPOUNDS.

For a partial list of these, see p. 290.

30

INDEX TO MINERALS.

INDEX.

Lightning Source UK Ltd.
Milton Keynes UK
UKHW012145150219
337363UK00004B/452/P